SpringerBriefs in Applied Sciences and Technology

SpringerBriefs present concise summaries of cutting-edge research and practical applications across a wide spectrum of fields. Featuring compact volumes of 50 to 125 pages, the series covers a range of content from professional to academic.

Typical publications can be:

- A timely report of state-of-the art methods
- An introduction to or a manual for the application of mathematical or computer techniques
- A bridge between new research results, as published in journal articles
- A snapshot of a hot or emerging topic
- An in-depth case study
- A presentation of core concepts that students must understand in order to make independent contributions

SpringerBriefs are characterized by fast, global electronic dissemination, standard publishing contracts, standardized manuscript preparation and formatting guidelines, and expedited production schedules.

On the one hand, **SpringerBriefs in Applied Sciences and Technology** are devoted to the publication of fundamentals and applications within the different classical engineering disciplines as well as in interdisciplinary fields that recently emerged between these areas. On the other hand, as the boundary separating fundamental research and applied technology is more and more dissolving, this series is particularly open to trans-disciplinary topics between fundamental science and engineering.

Indexed by EI-Compendex, SCOPUS and Springerlink.

More information about this series at https://link.springer.com/bookseries/8884

Pitri Bhakta Adhikari · Aashutosh Adhikari

Lightning Discharges

Formation, Terminologies, Measurements
and Theory

 Springer

Pitri Bhakta Adhikari
Department of Physics, Tri-Chandra M.
College
Tribhuvan University
Kathmandu, Nepal

Aashutosh Adhikari
Pulchouk Engineering College
Tribhuvan University
Kathmandu, Nepal

ISSN 2191-530X   ISSN 2191-5318 (electronic)
SpringerBriefs in Applied Sciences and Technology
ISBN 978-981-19-1925-1   ISBN 978-981-19-1926-8 (eBook)
https://doi.org/10.1007/978-981-19-1926-8

This Springer imprint is published by the registered company Springer Nature Singapore Pte Ltd.
The registered company address is: 152 Beach Road, #21-01/04 Gateway East, Singapore 189721,
Singapore

Preface

Benjamin Franklin, in 1752, speculated and verified the electrical nature of lightning. It took more than two centuries for scientists to carry out extensive research on lightning discharges since then. At present, more than hundreds of researchers from several geographical locations are engaged in investigating the nature of the lightning discharge. However, the complete phenomenon of lightning discharges is not well understood until today. Although the thunderstorms were observed to produce the most common lightning activities such as cloud flashes, ground flashes (both positive and negative), and air discharges, some unusual events were observed in hills and mountains. The unusual events were observed to have occurred quite frequently and generally produce positive electric field waveforms preceded by the opposite polarity leader-type pulse with a duration of microseconds. The cloud discharge occurs mostly from a negative charge center's lower and upper boundaries, and it bridges the two charge centers, upper positive charge and main negative charge.

Most researchers have researched the downward negative lightning discharges, which occur about 90% or more. It is generally thought that the lower positive charge region enhances the electric field at the bottom of the main negative charge region and thereby facilitates launching the negative charge toward the ground. The hills and mountains may influence the lightning activity leading to a greater number of ground flashes. Moreover, the density of positive ground flashes in hills and mountain regions has been relatively higher. Positive ground flashes are generally more hazardous as they constitute current of higher magnitude. To better understand the lightning phenomena, electric fields generated by lightning were measured and analysed. The majority of the positive ground flashes were found to be single-stroke ones.

In contrast, the average number of strokes per flash is 1.1, with a maximum value of 4 in the Himalayan mountainous region. Some of the positive cloud-to-ground lightning flashes were found to consist of return strokes with multiple peaks, called doubly peaked strokes. Besides this type of lightning, rocket lightning, ribbon lightning, bead lightning, ball lightning, etc. also occurred in nature.

Kathmandu, Nepal   Pitri Bhakta Adhikari
                                                         Aashutosh Adhikari

Contents

Chapter 1
Introduction

1.1 Overview

When sunlight provides sufficient amount of heat, it warms the upper surface of the water as well as the surface of the ground. The air in contact with the ground surface also warms up along with it. This air, on the surface of the ground, after being heated expands and becomes lighter than the surrounding air. As it goes up, the surrounding atmospheric pressure decreases and it expands further. As it goes up, the temperature decreases and the ability of air to hold the moisture (called humidity) decreases. As the temperature decreases when it moves upward, the moisture in the air condenses and forms water drops. Hence, by this process, thunderstorm clouds or cumulonimbus develop in the atmosphere (Rakov & Uman, 2003).

These water drops freeze and form tiny ice crystals, snow crystals, graupel particle, super-cooled water droplets, etc. in the cloud and still remain as water drops in the form of liquid even below 0 °C. They are called super-cooled water droplets. With the increase in height of the cloud, the water vapor changes in different forms as snow crystals, ice crystals, graupel particle, super-cooled water droplets, etc. These small ice particles, snow crystals, ice crystals, collide to increase their size and again collide with super-cooled water droplets to form a graupel particle. As the size of the graupel particles increases, they become heavier and start to move down. As the temperature difference is higher between the two layers of the cloud, they move faster. Due to the random motion of graupel particles, snow crystals, ice crystals, water molecules in the cloud, the electrification phenomena occurs by the process of friction and mutual contact between them. This produces large amount of charge in the cloud(s). The diagram of the charge structure of thunderstorm is the tripole standard model which contains the main negative charge at the center of the cloud, main positive charge above the negative charge, and a lower positive charge as a pocket charge on the bottom part of the cloud (Rakov & Uman, 2003; Uman, 2001). The structure of the cloud is shown in Fig. 1.1.

P. B. Adhikari and A. Adhikari, *Lightning Discharges*,
SpringerBriefs in Applied Sciences and Technology,
https://doi.org/10.1007/978-981-19-1926-8_1

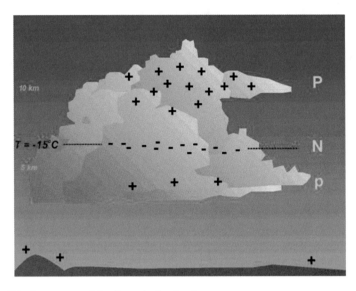

Fig. 1.1 Tripole structure of the charge in the cloud

The main charging area in a thunderstorm occurs in the central part of the storm where air is moving upward rapidly called updraft, and the main negative charge center is found in a relatively narrow temperature range from -10 °C to -25 °C, regardless of the height of the ground below this storm in which the cloud contains the super-cooled water droplet, ice crystal, etc. as mentioned by Malan (1963); Uman (2001); Rakov and Uman (2003); Cooray (2015). These ingredients, super-cooled water droplet, and ice crystal play important roles for charging mechanism of the position of charges in the cloud (Mac Gorman & Rust, 1998). At this position, the combination of temperature and rapid updraft, i.e., upward air movement, produces a mixture of below freezing small water droplets, called super-cooled water droplets, small ice crystals, and soft hail graupel particle. Super-cooled water droplets and very small ice crystals then move upward. At the same time, the larger and denser graupel particles tend to fall or be suspended in the rising air. As the particles are moving upward and heavier particles are moving downward, the differences in the movement of the precipitation cause collisions to occur. The updraft carries the positively charged ice crystals upward toward the top of the storm cloud. The larger and denser graupel with negative charge is either suspended in the middle of the thunderstorm cloud or falls toward the lower part of the storm (Cooray, 2015; Uman, 2001). The examples of different places at various heights having different temperatures are mentioned in Table 1.1. From these examples of Table 1.1, the structure of the cloud is generalized that the negative charge center lies in the central part of the cloud in the range of temperature -10 °C to -25 °C, main positive charge center lies above the negative charge center and below the temperature -25 °C and small positive charge pocket remains on the bottom part of the cloud as mentioned

Table 1.1 Examples of different places of various heights and temperatures

Researcher	Places	Heights	Charges	Temperatures
Krehbiel et al. (1983)	Florida	6–8	Negative	−10 °C to −25 °C
		8–15	Positive	Above −40 °C
	New Mexico	6–8	Negative	−10 °C to −25 °C
		8–15	Positive	Above −40 °C
	Japan	2–8	Negative	−10 °C to −25 °C
		8–15	Positive	Above −35 °C
Simpson and Robinson (1941)	England	6–15	Positive	Above −40 °C
		3–6	Negative	−10 °C to −25 °C
		1.5–3	Positive	0 °C to −5 °C
Winn et al. (1981)	New Mexico	4.8–5.8	Negative	−10 °C to −25 °C
		Above 10	Positive	Above −40 °C
Marshall and Winn (1982)	Japan	1–3	Negative	0 °C to −10 °C
Weber et al. (1982)	New Mexico	5.5–8	Negative	−5 °C to −20 °C
		Above 8	Positive	Below −25 °C

in different books and published articles such as Adhikari (2019); Cooray (2015); Krehbiel (1986); Krehbiel et al. (1983).

As mentioned by Cooray (2015), the rising water vapor and unsaturated air decreases the temperature at the rate of 1 °C per 100 m and after the saturation, the cooling rate becomes 0.6 °C per 100 m. Therefore, the stability of the atmosphere depends on the temperature, altitude, and the moisture content in the air. The variation in temperature as a function of height of atmosphere is shown in Fig. 1.2.

As shown in Fig. 1.2, as the height of the atmosphere above the surface of the earth increases, the temperature of the atmosphere decreases initially. This atmosphere at about 15–16 km from the surface of the earth is called troposphere, where the minimum temperature is about −55 °C. Again, when the height of atmosphere increases, the temperature remains nearly constant for a certain height up to about 40 km. This surface of the sphere is called stratosphere. The line of demarcation between the stratosphere and troposphere is called tropopause from which the temperature starts to increase. In stratosphere, the minimum temperature is about −20 °C. Again, this temperature is nearly constant and slightly decreasing in a region called mesosphere, in which the upper atmospheric lightning occurs. These upper atmospheric lightnings are also related to the lightning phenomena that occur in the troposphere. Hence, the lightning occurring in the atmosphere depends on the electrical characteristics of the atmosphere, i.e., the ion density, electron density, electrical conductivity, and the density of the gas of the atmosphere.

As mentioned in Table 1.1, different places of various heights have different temperature and different position of the negative and positive charge. As mentioned in Table 1.1, the temperature at the height of 5–8 kms is in the range of −10 °C to

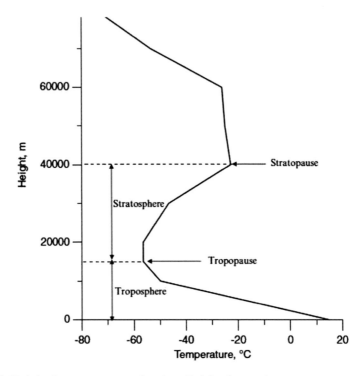

Fig. 1.2 Variation in temperature as a function of height of atmosphere

−25 °C. In this region, negative charge center resides and above that region on the upper atmosphere there is positive charge center.

The composition of water droplet and ice crystals forms clouds in the earth's atmosphere, which are commonly white in appearance. The cloud appears white because the liquid and solid particles have high wavelength relative to visible light. Air becomes super-saturated with water vapor; then, the super-saturated air is lifted up and cooled by adiabatic expansion. The lifted process is caused by heating of the air near the earth's surface, and also it is heated by sunlight directly. The warmed air is lifted up at a height in which the super-saturation condition is achieved. When the air is lifted up at 1000 m from the earth's surface, the temperature of this air becomes around 0 °C. Higher the air, lower is the temperature than 0 °C. At about 7000 m above the earth's surface, the temperature reaches about −24 °C or even less than that. The temperature at different height at different places should be different; it may depend on various factors. The actual temperature of the atmosphere measured from aeroplane can be taken as an example which is given in Table 1.2. From Table 1.2, we can say that if the height of the air is uplifted, the temperature of the air decreases.

Table 1.2 Temperature and height of the cloud measured directly inside the cloud

S. No.	Height of the cloud from earth's surface (m)	Temperature of the cloud (°C)
1	1000	0
2	1000	−1
3	1500	−3
4	2000	−3
5	2500	−4
6	3000	−6
7	3500	−6
8	4000	−8
9	4500	−12
10	5000	−13
11	5500	−17
12	6000	−19
13	6500	−23
14	7000	−24
15	7500	−27
16	8000	−29
17	8500	−39
18	9000	−50
19	9500	−62
20	10,000	−63
21	10,500	−65
22	11,000	−64
23	11,500	−65
24	12,000	−64

1.2 History of Lightning

1.2.1 Ancient to Middle Age

Lightning, fear and respect in mankind, from the ancient time of civilization, is very important in different religions and mythology (Uman, 1987). In 1964, Schonland reviewed the non-scientific views on lightning and thunder over the duration of 5000 years. The cloud to ground lightning in non-scientific literature is reported by "thunderbolt". He found the following facts at that time, which were related to the mythology about the lightning. The different mythology about lightning is shown in Fig. 1.3 in which the symbol are imagined to produce lightning, and provide punishment to the evil.

Fig. 1.3 Examples of different mythologies that are imagined to produce lightning, and provide punishment to the evil. **i** lightning dragon in Chinese mythology; **ii** and **iii** eagle emblem in one dollar and eagle produced lightning in American mythology; **iv** Zeus in Greek mythology; **v** King Indra carried thunderstorm in Indian mythology; **vi** Scandinavian mythology; and **vii** Mythology in Nordic culture

- Early statues of Buddha showed him carrying a thunderbolt with arrowheads at each end.
- In Egypt, the Typhon god hurled the thunderbolt.
- On Vedic books of India, the son of Heaven and Earth Indra carried thunderbolts in his chariot.
- Zarpenic, the lightning goddess, takes thunderbolts in each hand generating thunder with a whip.
- Lightning was viewed in ancient Greece as punishment sent by Zeus, the father of the Gods.
- The Chief God of the Romans, Jupiter or Jove, was thought to use thunderbolts not only warning against undesirable behavior but also retribution.
- In one dollar bill in USA, the eagle emblem of Jupiter represents the thunderbolt.
- In Rome, College of Augurs was charged for the location of the lightning.
- The fierce god of the Norsemen Thor produced lightning by hitting with a hammer on his anvil.
- Jupiter's approval: the term, "Thursday" was originated from Thor's day (meaning, lightning day). It is also referred to as Thors-day in Danish, Thunder day in German, and Jove's day in Italy.

- Meteorites, referred thunderstorm in Scandinavia, as a stone are broken by Thor's hammer, struck on his anvil.
- Flashing feather of a mystical thunderbird whose flapping wings produced thunderstorm and lightning in North American as well as in African mythology, which dives from cloud to earth.
- In Northern Syria, the god Teschup generated thunder.
- In Chinese mythology, the goddess Tien Mu, chief among the five ministry of thunderstorm, produced thunderbolt.
- In Russia, Buryats believed that lightning produced by their god and the god throws stones from the sky.

1.2.2 From Middle Ages to Benjamin Franklin

In the period of 33 years as mentioned in several books such as Uman (2001); Rakov and Uman (2003); Malan (1963), etc., the lightning struck 386 church steeples killing 103 bell ringers while doing their duties. In 1388, the campanile of St. Mark in Venice, Italy was damaged by lightning. In 1417, it was set on fire and destroyed again in 1489, and continuously in 1548, 1565, 1653, and 1745. After the era of Benjamin Franklin 1766, lightning rod has been used in this church and no further damages have been done since then. In 1718, 24 church towers in France were damaged by lightning, in 1769, the steeple of the church of St. Nazaire in Brescia, Italy was struck by lightning and killed three thousand people.

1.2.3 Era of Benjamin Franklin

Benjamin Franklin was the first scientist who proved that the lightning phenomenon is the electrical discharge phenomenon by doing the two separate experiments in various times. He did two famous experiments, the first on sentry-box experiment and second one is kite experiment in 1752; Thomas Francois D'Alibard performed the sentry-box experiment successfully as suggested by Benjamin Franklin. In this experiment, (i) sentry-box experiment which proves that the thunderclouds are electrified. Inside the sentry box, man holds an iron rod standing on the electrical stand and discharge occurs from other hand in the first diagram, but on the second diagram standing on ground holding insulating wax handle, there are sparks that occur between iron rod and a grounded wire. (ii) The kite experiment proves that the lightning phenomenon is only electrical discharge phenomenon adapted from Uman (2001) shown in Fig. 1.4.

From this experiment, it was successfully proved that lightning phenomenon is nothing but electrical discharge. After the successful completion, the same experiment was repeated in France, and again in England and Belgium. But, in July 1753, a Swedish physicist G. W. Richmann, working in Russia, was killed by a direct lightning strike. From the sentry-box experiment and the famous kite experiment, it

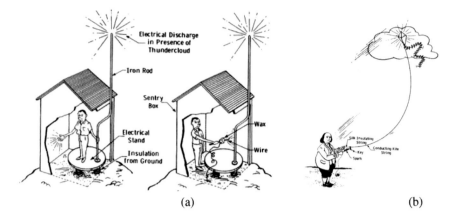

Fig. 1.4 Benjamin Franklin, **a** sentry-box experiment and **b** kite experiment

was proved that the lightning phenomenon is electrical discharge. He invented the lightning rod, for the electrical discharge phenomena which can be used to prevent the buildings. Thereafter, the lightning rods were used in different church, temples, buildings, which is the primary tool of protecting structures against the lightning. Hence, Franklin's invention is very useful till now.

1.2.4 *Modern Time*

After the era of Benjamin Franklin, there wasn't any significant progress up to one and half century. Lightning research has significant progress since photography and spectroscopy measurements became a tool for research. According to Uman (2001), lightning photography is a diagnostic tool to take the photo of lightning phenomena in lightning research. In 1889, Hoffert in England received the photo of lightning discharge to ground. Similarly, Weber in 1889, Walter in 1902, 1903, 1910, 1912, and 1918 in Germany, AND Larsen in 1905 in the USA also took photographs. Boys in 1926 in England invented the double-lens streak camera, due to which the tools of photography have been more successful in lightning research. Investigators used time-resolved photographic techniques in the USA, England, and Germany to identify individual strokes, leader steps, etc. of cloud to ground discharges. The double-lens streak camera was invented in 1900 by Boys in England and was used for the cloud to ground lightning by different researchers such as Schonland (1964); Malan (1963), etc. in South Africa after 1930. Pockels in 1897, 1898, and 1900 in Germany made the first lightning current measurements, and due to this, he analysed the residual magnetic field. After Wilson (in 1916) in England, modern lightning research was started by measuring the electric field measurement system. He was the first to use electric field measurement system and estimate the charge structure in the thunderstorm and the charges involved in lightning discharges. He received a Nobel Prize for the invention of the cloud chamber to track high-energy particles.

Beside these measurements, there are spectrometer measurements, thunder measurements, etc. The remote ground-based electric field measurement system was used by Wilson to determine the charge structure in the thundercloud and the charge involved in the lightning discharge phenomena. This method is frequently used by several researchers from the period of Wilson till now. Initially, the combination of two methods, high-speed photographic and electric field measurement, was used by Schonland in South Africa. In Switzerland, Berger (1977) measured the lightning currents. Since 1970, more researchers have actively participated in the field of lightning. During this time, new techniques of data acquisition and analysis have been developed with the help of sophisticated instruments and development of computers for the storage of data. During the last two decades, the rocket-wire triggered lightning has been used for research purpose to study various aspects of lightning discharge. Nowadays, some researchers have been involved to study the upper atmosphere such as red sprite, halos, jets, elves, etc. and the high energy phenomena such as electrons, X-rays, gamma rays including Terrestrial Gamma Ray Flashes (TGFs) observed in orbit of satellite (Dwyer & Uman, 2014). They are produced by lightning discharge phenomena. The observation due to analysis of the waveform produced by the lightning discharge phenomena still has many mysteries.

References

Adhikari, P. B. (2019). Variation of atmospheric temperature with height in the phenomena of lightning waveforms. *World Journal of Applied Physics, 4*(4), 46–50.

Berger, K. (1977). *The earth flash, in lightning, physics of lightning*. Academic Press.

Cooray, V. (2015). *An introduction to lightning*. Springer.

Dwyer, J. R., & Uman, M. A. (2014). The physics of lightning. *Physics Reports, 534*, 147–241.

Krehbiel, M. P. Brook, R., Lhermitte, R. L., & Lennon, C. L. (1983). Lightning charge structure of thunderstorm. In: *Proceedings of conference on atmospheric electricity* (pp. 408–411). Hampton, Virginia.

Krehbiel, P. R. (1986). *The electrical structure of thunderstorm in the earth's electrical environment*. National Academy Press.

Mac Gorman, D., & Rust, W. (1998). *The electrical nature of storms*. Oxford University Press.

Malan, D. J. (1963). *Physics of lightning*. English University Press.

Marshall, T. C., & Winn, W. P. (1982). Measurements of charged precipitation in a New Mexico thunderstorm: Lower positive charge centers. *Journal of Geophysical Research, 87*, 7141–7157.

Rakov, V. A., & Uman, M. A. (2003). *Lightning: Physics and effects*. Cambridge University Press.

Schonland, B. F. J. (1964). *The flight of thunderbolts*, 2nd edition. Clarendon Press.

Simpson, G. C., & Robinson, G. D. (1941). The distribution of electricity in the thunderclouds. *Proceedings in Research Society, London series A, 117*.

Uman, M. A. (1987). *The lightning discharge* (p. 377). Academic Press.

Uman, M. A. (2001). *The lightning discharge* (p. 377). Dover Publications.

Weber, M. E., Christian, H. J., Few, A. A., & Stewart, M. F. (1982). A thundercloud electric field sounding: Charge distribution and lightning. *Journal of Geophysical Research, 87*, 7158–7169.

Winn, W. P., Moore, C. B., & Holmes, C. R. (1981). Electric field structure in an active part of a small, isolated thundercloud. *Journal of Geophysical Research, 86*.

Chapter 2
Lightning Terminology

2.1 Cloud and Its Formation, Cumulonimbus, Thunderstorm

The primary source of lightning is cumulonimbus, but every cumulonimbus does not produce lightning discharges. The cumulonimbus that produces lightning discharge is more appropriately called a thunderstorm cloud or only a thunderstorm. The thunderstorm cloud develops from the water droplet and vapor produced in the atmosphere from the earth's surface. From the sun's heat, water on the earth's surface changes its phase. The process of formation of cumulonimbus and thunderstorms is already mentioned in Sect. 2.1. In the process of lightning discharge phenomena, colossal energy (such as light, heat, sound, electromagnetic radiations, etc.) is released in different forms. Several electromagnetic radiations are produced due to the lightning discharge of varying frequencies. It may be low frequency to the higher order of frequency up to a few GHz. So, it is an extremely complex electrical discharge phenomenon. Hence, a complete theory of it does not exist to this date although it is one of the most common natural activities (Adhikari, 2019).

2.2 Flash Density

The flash density can be defined as the number of lightning flashes per unit time per unit area. The instrument's flash counters measure the flash density. The flash density over a month represents the total number of lightning flashes over this one month per unit area (km^2). If we take the number of lightning flashes over a year, the flash density has the unit in flash per year per km^2. There are several flash counters which are used to count the lightning flash. The CIGRE-500 Hz counter and CIGRE-10 kHz counter are widely used to count the lighting flash as described by Uman (2001). Different types of lightning flash counters are characterized by using different filters

P. B. Adhikari and A. Adhikari, *Lightning Discharges*,
SpringerBriefs in Applied Sciences and Technology,
https://doi.org/10.1007/978-981-19-1926-8_2

Table 2.1 Flash density at different locations

Place	Researcher	Flash density
Australia	Prentice and Mackerras (1969)	5 km^{-2} yr^{-1} TFD
	Mackerras (1978)	1.2 km^{-2} yr^{-1} GFD
	Prentice (1977)	
Scandinavia	Muller-Hillebrand (1965)	
South Africa	Anderson (1980)	
	Anderson and Eriksson (1980)	
Norway, Sweden, and Finland	Prentice and Mackerras (1969) Mackerras (1978)	0.2–3 km^{-2} yr^{-1} GFD depending on locations
South Africa	Prentice (1977) Anderson (1980), etc.	0.1–12 km^{-2} yr^{-1} GFD depending on locations
Florida	Lopez and Holle (1986)	8 km^{-2} yr^{-1} GFD
Colorado	Lopez and Holle (1986)	7 km^{-2} yr^{-1} GFD
Tampa-Bay, Florida	Darveniza and Uman (1984)	7–17 km^{-2} yr^{-1} GFD
Canada	Crozier et al. (1988)	1.6 km^{-2} yr^{-1} GFD in 1982
		2.4 km^{-2} yr^{-1} GFD in 1983

and triggering different thresholds. For different strokes, the electric field intensity also varies and depends on the distance of the lightning. So, the counter may fail to register the less effective event, and it can capture the high effective event of far distance. Henceforth, the reported values vary considerably, and some examples of different data are mentioned in Table 2.1.

2.3 Charge Structure in the Cloud

As described by most scientists, the charge structure in the cloud is similar. The positive charge region is at the top of the cloud, and the negative charge region is at the center of the cloud, forming a dipole in the cloud as the fundamental dipole structure. However, later on, the researchers found some positive pocket charge remained at the bottom part of the thundercloud and explained about the tripole charge structure of the cloud (Rakov & Uman, 2003; Uman, 2001). The tripole charge structure of the cloud is shown in Fig. 1.1. Thundercloud generally contains two main charge centers, a positive charge at the top of the cloud and the negative charge just below the positive charge, and a small pocket of positive charge located at the cloud base. The charge center's location appears to be determined by temperature and not by the height of the cloud above the ground. The main negative charge center lies within the temperature between −10 °C and −25 °C, and the main positive charge center lies some km above the negative charge. The main positive charge center lies at a very low temperature lower than $-25\,°C$. There is a small pocket of positive charge below the main negative charge center, close to the freezing level, that appears to be associated with the precipitation shaft.

If the charge transfers to the ground from the cloud, then ground flash occurs. The ground flash brings the positive charge from cloud to earth, which is called positive ground flash, and the flash that brings negative charge from cloud to earth is called negative ground flash. Again, with the increase in the number of charges, the electric field also increases, and the maximum electric field just before breaking the molecules is called the medium's dielectric strength. If the electric field is greater than 30 kV/cm, then the air (initially insulator) becomes a conductor, and there is an electric breakdown of the air. A ground flash is initiated by an electrical breakdown process in the cloud called preliminary breakdown. This process leads to creating a column of charge called the stepped leader that travels from cloud to ground in a stepped manner.

2.4 Preliminary Breakdown

The lightning initiation process inside the thunderclouds is mysterious. No one knows the process, which is quite impressive just above our surroundings in the atmosphere. In the atmosphere within the range of about 8–10 km, the lightning initiation process occurs inside the thundercloud. Hence, many researchers have given the idea about the lightning initiation process inside the thunderstorm cloud. To start this process, the charges are attracted toward the unlike charges, and they initiate by moving inside the thundercloud. When the charge moves in any direction, it searches for the path to move. During this process, it travels and suffers various processes and moves here and there. The negative charges at the center of the cloud are moving in random motion. Also, the positive pocket charges are just at the bottom part of the cloud. Thus, the negative charges are attracted toward it. If there are excess negative charges, they move further along this direction, but if the positive pocket charges are sufficient, the moving negative charges disappear. This means that they stop the initiation of the negative charges. If the negative charges move further, they go outside the cloud and travel up to the air. Sometimes, these charges can move even further and may even reach the ground (Nag & Rakov, 2009).

The initiation of the charge moving process leads to creating a propagating hot leader channel inside the cloud. This hot leader channel inside the cloud is called an initial breakdown process or preliminary breakdown process. This process is the initiation process of the lightning phenomena. To understand the lightning initiation process, we should know the microphysical discharge process, which produces the leader channel. The large electric field produced inside the thundercloud produces lightning discharge phenomena.

The leader channel, which is created inside the thundercloud, called streamers, propagates and finally forms lightning. The streamers are very weak and move at slow speed. These streamers are isolated electrically from the point of origin, and if

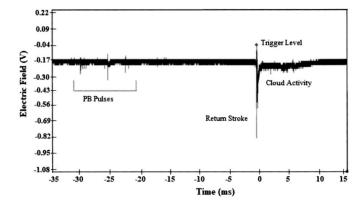

Fig. 2.1 An example of a preliminary breakdown pulses adopted from Adhikari (2019)

they take more charge and propagate faster, they form a stronger field. They increase the conductivity and travel more distance. On traveling this way, it forms the branches allowing multiple channels (Clarence & Malan, 1957). Researchers (such as, Beasley et al., 1982; Clarence & Malan, 1957; Krider et al., 1975; Weidman & Krider, 1979) investigated and indicated that on the beginning of stepped leader or at the end of preliminary breakdown, there are relatively large bipolar pulses with the system operating from very low frequency to very high frequency. To initiate the lightning discharge, the electric field inside the thunderclouds should be larger than 2 × 106 V/m (Uman, 2001). The example of preliminary breakdown pulses is shown in the following Fig. 2.1 (Adhikari, 2019).

2.5   Stepped Leader

Schonland and his coworker in South Africa determined the stepped leader in 1930 by using streak photographic measurements. After that, many other researchers have studied about the stepped leader, such as Krider et al. (1975); Cooray and Lundquist (1985); Beasley et al. (1982); Uman (2001); Rakov and Uman (2003), etc. All of them studied the stepped leader, which propagates step by step, as a stepping process, and each step has an average length of about 40–50 m with a range between 10–200 m. The pause time between the steps is 37–124 μs (Schonland, 1956). The step extends the length of the leader channel and further moves, which is one of the critical processes of lightning formation. The stepped leader initially moves at a small distance from the charge of the cloud toward the earth. Then, slowly, the leader's speed increases and moves faster. If the value of the speed of the stepped leader is in the order of 105 m/s, it is called a stepped leader, and these leaders are shorter in length and much luminous than the stepped leader. These leaders move with long steps, bright and high average speed in the order of 106 m/s (Beasley et al.,

1982). Examples of the stepped leader before the return stroke are shown in Fig. 2.2 (Adhikari et al., 2016) (Table 2.2).

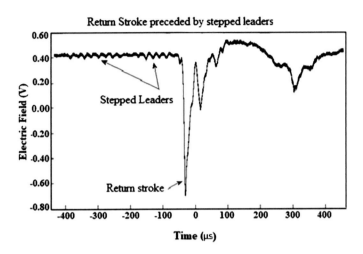

Fig. 2.2   An example of positive return stroke preceded by stepped leaders

Table. 2.2   Velocity of stepped leader at different places by different researcher

Researcher	Location	Velocity of leader ($\times 10^5$ m/s)	Distance	No. of leaders
Schnoland (1956)	South Africa	0.8–8	2–3 km	60
Berger and Vogelsanger (1966)	Switzerland	0.9–4.4	1–3 km	14
		1.9–2.2	100 m	4
Chen et al. (1999)	Australia	4.5–11	367–1620 m	1
	China	4.9–5.8	33–102 m	1
Thomson (1985)	Florida	1.3–19	600 m–2 km	10
Proctor et al. (1988)	South Africa	0.3–4.2	–	66
Beasley et al. (1982)	Florida	8–39	100 m	3
Nagai et al. (1982)	Japan	1.7–4.7	Over 1 km	–
Shao et al. (1995)	Florida	2	–	–
Isikawa (1961)	Japan	3.1	–	13
		3.2		14
Rakov and Uman(1990)	Florida	2	–	–

2.6 Attachment Process

By moving the leader step by step to the ground, when the leader approaches any conducting object, the electric field produced by the charge can be enhanced by the objects (such as a transmission line tower, a tree, or an aircraft in flight, etc.) and then discharge phenomenon occurs. The leader's process, which just touches and the discharge phenomena occurs, is called the attachment process. This attachment process plays a significant role in designing lightning protection, overhead ground wire protection (Armstrong & Whitehead, 1968). Among the parameters for lightning protection, the striking distance, i.e., the distance between the object to the strike and the tip of the downward-moving leader at the instant, is significant. This striking distance plays a vital role in designing the protection system from lightning. It is assumed that the striking point lies in between the tip of the last step leader and the object, which is called the lightning stroke point.

Golde (1945, 1947, 1967) reviewed the phenomena of the attachment process related to the lightning stroke to the object attached to the ground by calculating the resulting fields on the ground due to the charge on the leader channel. Then, he derived the relations between the leader charge and the striking distance. Hence, the striking distance is related to the magnitude of the following return stroke's peak current. Since the charge of the stepped leader may be spread over the branching, the relation of the current and the striking distance is not clear practically (Eriksson, 1978). However, there is a relationship between the peak current of the return stroke at the striking point and the total charge transfer to the striking object, as mentioned in Uman (2001). Golde (1977) gave the relation by analyzing the several theoretical relations between the striking distance and the peak current as $d = 10 \times I \, 0.65$, where d is the striking distance in meters and I is the peak current in the return stroke measured in kA units.

2.7 Return Stroke

The amount of charge the cloud lowers to the earth (ground) is called return stroke. This process is best to understand the makeup of a flash to earth. The return stroke that lowers the positive charge to earth is discussed less than the lower negative charge to earth. If the negative charge of the cloud transfers to the earth, the amount of negative charge decreases in the cloud, and it is called a negative return stroke. Similarly, if the positive charge transfers to the ground, the amount of positive charge decreases in the cloud, and it is called a positive return stroke. Most of the researchers have studied the negative return stroke, which generally occurs about 90% in the atmosphere (Rakov & Uman, 2003; Uman, 2001). The process of return stroke is the optically brightest lightning process, and it is easily identified as the main event in the electromagnetic signature.

To measure the return stroke, there are various methods, but mainly they are categorized into four types, which are:

1. Electric and magnetic field measurement
2. Current measurement
3. Speed measurement
4. Luminosity and optical spectra.

These methods are discussed in detail below.

2.7.1 Study About the Return Stroke by Using Electric and Magnetic Field Measurement

By using the electric field measurement, which is the third step measurement of lightning phenomena, many researchers and scientists have known about the nature of lightning phenomena and the charge structure in the cloud. Since it was started by Wilson in England in the second decade of the nineteenth century, other researchers have followed this process and published many research articles and added to the knowledge of the mystery of the charge structure in the cloud and the nature of lightning phenomena. Many researchers such as M. Brook, N. Kitagawa, P. R. Krehbiel, Y. T. Lin, J. A. Tiller, R. D. Brantley, etc. have published research papers about the return stroke by measuring the electric field variations on a millisecond time scale, but R. J. Fisher, Y. T. Lin, J. A. Tiller, R. D. Brantley, C. D. Weidman, E. P. Krider, M. A. Uman, W. H. Beasley, etc. measured the vertical electric field and horizontal magnetic field variation of return stroke on the microsecond time scale as mentioned in Uman (2001).

G. I. Serhan, C. D. Weidman, E. P. Krider, M. A. Uman, J. Preta, D. G. Childers, J. C. Willett, J. C. Baisley, and so on have published research articles about the frequency spectra of the measured electromagnetic radiation fields. Within a few kilometers distance of the flash, the electric field of strokes observed are dominated by the electrostatic field component until few tens of microseconds. The magnetostatic component dominates the total magnetic fields at closer distances of the lightning flash (Lin et al., 1979). In the total electric field which is a combination of the electrostatic field, induction field, and radiation field, the radiation field is in the opposite direction of the combination of electrostatic and induction fields. At closer distance, the radiation field is dominant but far away, the induction field and electrostatic field dominate the radiation field (Thottappillil et al., 1992, 2001). Krider and Guo (1983); Peckham et al. (1984), etc. examined experimentally that keeping the threshold triggered value for the electric field according to the distance, the mean of electric field for the first stroke ranged from 6 to 8 V/m and that for subsequent strokes ranged from 4 to 6 V/m.

2.7.2 Study About the Return Stroke by Using Parameters Derived from Channel Base Current Measurements

Pockels in Germany started using the current measurement of lightning phenomena, which is the second step of measurement, by doing experiments in 1897, 1898 and 1900. Some features of lightning phenomena were known by measuring current. Berger and coworkers in Switzerland measured the current at the base of stroke channel of the lightning and reviewed and published research articles (Berger, 1977; Berger & Vogelsanger, 1966; Berger et al., 1975). Berger et al. (1975) measured the current, minimum peak value in the distribution as 2 kA and other parameters, downward CG lightning derived from the channel base measurement as given in the Table 2.3.

The current of the lightning stroke channel was measured by using the voltage induced in resistive shunt which is used in the top of towers. This type of process to measure the voltage on the resistive shunt and the current measurements were made by Garbagnati and his coworkers in Italy (Garbagnati et al., 1978). Eriksson (1978) in

Table. 2.3 Lightning current parameter adapted from Berger et al. (1975)

Parameters		No. of events	Percentage of cases exceeding tabulated value		
			95%	50%	5%
Peak current (kA) (Minimum 2 kA)	Negative first strokes	101	14	30	80
	Negative subsequent strokes	135	4.6	12	30
	Positive first strokes	26	4.6	35	250
Charge (Coulomb)	Negative first strokes	93	1.1	5.2	24
	Negative subsequent strokes	122	0.2	1.4	11
	Negative flashes	94	1.3	7.5	40
	Positive flashes	26	20	80	350
Stroke duration (μs) (2 kA to half value)	Negative first strokes	90	30	75	200
	Negative subsequent strokes	115	6.5	32	140
	Positive first strokes	16	25	230	2000
Time interval between negative strokes (ms)		133	7	33	150
Flash duration (ms)	Negative (single-stroke flash)	94	0.15	13	1100
	Negative (multiple-stroke flashes)	39	31	180	900
	Positive (single-stroke flash)	24	14	85	500

South Africa also used the same type of measurement but very fast current rise time was observed. The stepped leader typically is 1 μs in duration with a time interval of 20–50 μs between steps and the current in between 100 and 1000 A, the peak current 1 kA. The current of the return stroke in average is 30 kA, and maximum peak current is about 300 kA with the temperature peak about 30,000 K. The current of the subsequent return stroke is also measured whose peak value is 10–15 kA in less than 1 μs, then decays in some microseconds. But Berger and Vogelsanger (1965) stated that the rise time from bottom to peak should not be less than 1 μs. There are some other researchers such as S. Szpor, M. A. Sargent, F. Popolansky, D. G. McCann, etc. in the USA; B. N. Gorin, A. V. Shkilev, etc. in Russia; A. M. Hussein, W. Janischewskyj, etc. in Canada; O. Beierl, F. Fuchs, etc. in Germany; K. Miyake, C. Guto, etc. in Japan; E. Montandon, etc. in Switzerland; G. Diendorfer, etc. in Austria; M. Lacerda, etc. in Brazil; etc. who made the direct measurement of the current of the natural lightning return stroke.

2.7.3 Speed of Return Stroke

By measuring the speed of the return stroke of the lightning wave-front, such as Jordan and Uman (1983), they studied the luminosity of the stroke which varies with height. The measurement of return stroke speed was started by Schonland and Collens (1934), Schonland et al. (1935), in South Africa. They used different cameras such as Boys camera, two-lens streak camera, etc. to study about the speed of the return stroke, which was found to be 1×10^8 m/s and decreased at the branch point. Other researchers, such as J. S. Boyle, R. E. Orville, V. P. Idone, and so on, published on the speed of return stroke for natural lightning and P. Hubert, V. P. Idone, G. Mouget, etc. and also studied about the speed of return strokes of artificial lightning.

2.7.4 Optical Spectra and Luminosity

Hoffert in England and Weber and Walter in Germany used the first method to know the nature of lightning by photographic process in 1889. After that, Benjamin Franklin proved that lightning is an electrical discharge phenomenon by conducting two experiments: Sentry-box experiment and kite experiment, which started the research about lightning. This measurement of lightning photography is the first method, and to process this method was very easy by inventing special Boys camera in 1926 (Invented by Boys) (Uman, 2001). There has been some research about lightning spectroscopy, published by Uman and McLain (1969); Idone and Orville (1982); Jordan and Uman (1983); Idone and Orville (1985), from these spectra, other physical properties of lightning discharges such as temperature, pressure, particle density, etc. have been determined. Within the first 10 μs of the discharges, R. E. Orville found the temperature of the peak 10 return stroke in the order of 30,000 K, then within the range 10–20 μs, in between 30,000 and 20,000 K and after 20 μs, the temperature below 20,000 K (Orville, 1968a, b, c). By using this process, the time resolved spectrum found that the initial temperature of the stroke varied more rapidly than the time resolution. Similar cases were found in the stepped leader as well (Orville, 1968d) and for the dart leaders also (Orville, 1975).

R. E. Orville used the technique of measuring and comparing the stark width of the H-line of the H atom, which was found to be independent of the population of the atomic energy band. The electron density of the first 5 μs is about 10^{18} cm^3, and the pressure is 8 atm (Orville, 1968a, b, c). Some researchers G. E. Barasch, T. R. Connor, R. E. Orville, B. F. J. Schonland, etc. have carried out extensive spectroscopic study of lightning. Also, E. P. Krider, D. Mackerras, C. Guo, etc. have measured the luminosity of the return stroke with wide-field photo-electric system (Uman, 2001).

2.8 Dart Leader

After the first return stroke in the cloud to ground lightning discharge, the leader which is tens of meters in length and propagating toward the earth without branching is called a dart leader. Schonland and Collens (1934) measured the length of 9 dart leaders and found to be 54 m on average with a range of 25–112 m. Orville and Idone (1982) also proved this result by doing similar analysis of 11 dart leaders and proved that dart leader speed and dart length are perfectly correlated. The more energetic leaders should be faster, and it also provides more heating channel. The summary of speed of the dart leaders and sample size of the dart leaders is mentioned in Table 2.4.

Table. 2.4 Summary of sample size and speed of the dart leaders

Researchers	Location	Sample size	Mean speed of dart leaders (10^6) m/s
Schnoland (1956)	South Africa	55	5.5
McEachron (1939)	USA	17	11
Brook and Kitagawa (1964)	New Mexico	103	9.7
Berger (1967)	Switzerland	80	9.0
Hubert and Mouget (1981)	France	10	11
Orville and Idone (1982)	Florida and New Mexico	21	9
Jordan et al. (1992)	Florida	11	14
Mach and Rust (1997)	Oklahoma, Alabama and Florida	17	19

Table. 2.5 Comparison between the speed and length of the drat leaders and the time interval of dart leaders

Speed of dart leader (10^6 m/s)	Length of steps (m)	Time interval between the dart stepped leader (μs)
0.48	12	7.4
1.0	7.4	7.4
1.1	10	9.0
1.2	9	7.4
1.7	25	15.0
1.7	13	7.8

Orville and Idone (1982) found the lower limit $(1–3) \times 10^6$ m/s and the upper limit $(21–23) \times 10^6$ m/s of the speed of the dart leader. Schonland et al. (1935) proved that if the subsequent strokes have more luminosity, then the strokes have longer time interval. Schonland (1956) found the average speed, average step length, and the time interval between steps preceding the second return stroke of a multiple-stroke flash, which are mentioned in Table 2.5.

Barasch (1970) reported that emission of dart leader is always less than the emission of return stroke. Orville (1975) proved that the intensity of dart leader is 10 times less than the intensity of the return stroke, and the peak temperature becomes more than 20,000 K. The radiation produced by dart stepped leader was researched by various researchers such as Brook and Kitagawa (1964), Proctor (1971), Wang et al. (1999). Rakov and Uman (1990); Davis (1999), etc. used the measurement of the rate of change of electric field in Florida and found the inter-stroke interval of 61 ms in multistroke channel. This report also supports the report given by Guo and Krider (1982). Krider et al. (1977) found that the mean time interval of the dart stepped leader of the electric field pulses is 6.5 μs in Florida and 7.8 μs in Arizona.

2.9 Continuing Current

The return stroke in a cloud to ground flash has the peak current in the order of 100 kA, and the time for this is in the order of 100 μs. After the return stroke, the current about 10–100 A flows in the same channel of the return stroke in the time interval of tens to hundreds of milliseconds, which is called continuing current. This current is also called intermediate current which was first observed by Hagenguth and Anderson (1952). Some current components are below the duration of 40 ms (Livingston & Krider, 1978), and some researchers have shown the current component of duration above than 40 ms (Brook et al., 1962; Fuquay et al., 1967; Kitagawa et al., 1962). Nakahori et al. (1982) explained that the continuing current flows through the channel with the thermal effects, and it can make holes in a metal sheet of aircraft as well. Shindo and Uman (1989); Rakov and Uman (1990), etc. have studied about the initiation of long continuing currents and found the initial peak of the electric field to be smaller than the regular strokes. Fisher et al. (1993) in Florida and Alabama studied about the triggered lightning of the duration of 10 ms of continuing current.

2.9.1 M-Component

Malan and Collens (1937) originally identified this lightning process called M-component, which is the luminous continuing current channel. Due to the rapid electric field variation in the channel, the luminosity occurring in itself is called M-electric field change. Only the relation between the current and M-components was observed by Fisher et al. (1993) but Rakov et al. (1998) observed the relation between the current and the magnetic and electric field of M-components. Shao (1993); Shao et al. (1995); Mazur et al. (1995); Rakov et al. (2001) used the VHF images of M-components' channels of the lightning flash and characterization of M-components with critical review by some researcher such as Thottappillil, Rakov, Jordan, etc. as in their papers (Thottappillil et al., 1990, 1995; Rakov et al., 1992, 2001; Jordan et al., 1995).

In Fig. 2.3, the hook-shaped form represents the M-component field changes of the lightning flash. The field change of M-component is generally one-fifth to one hundredth of the field change of the return stroke wave-front propagation. This M-field changes occur in both short and long continuing current observed by Kitagawa et al. (1962). Without involving the leader process, a momentary current increases in the lightning channel to ground described the M-component by Kitagawa et al. (1962).

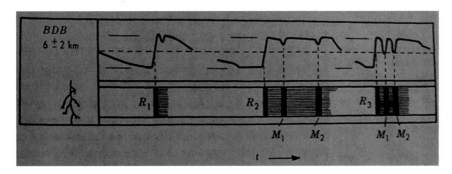

Fig. 2.3 Electric field change for the M-components observed near the lightning discharge adopted from Uman (2001)

2.9.2   K-Changes

Between the two strokes of a multiple-stroke flash, the small, rapid electric field change occurs, found by N. Kitagawa, which is called K-changes. During continuing luminosity, the M-field changes occurring at a distance is observed which are called K-changes (Kitagawa, 1957; Kitagawa & Brook, 1960; Kitagawa et al., 1962). Kitagawa et al. (1962) reported that the K-changes are associated with the luminous waves which fail to reach the ground, or with the luminosity in the cloud. This suggests that the streamers in the cloud can change the field which may be related to the K-changes in the intra-cloud lightning discharge.

2.10   J and K Process

The "J" or Junction process can be defined as the redistribution of the cloud charge between the two strokes in the multiple-stroke lightning flash. It can also be defined as the redistribution of the charge in the cloud after the first return stroke in the cloud to ground lightning discharge. This phenomenon of redistribution of charge takes about tens of milliseconds, and the electric field change in the wave shape is almost steady. The electric field change may either be in the same direction or in the opposite direction of the field change in the return stroke. Malan (1965) mentioned that in the cloud to ground flashes within a few kilometers, the J-field change between the strokes are always negative but beyond 5 km, it can be either positive or negative, i.e., at close distance, the J-field change is negative and the J-field changes to mixed polarity with the increase in distance of the lightning discharge.

The electric field change in J process is smaller than the field change due to continuing current in the lightning discharge. Brook et al. (1962), Berger (1967), Krehbiel et al. (1979), etc. had measured the electric field at different stations. Various photographic and visual observations by Brantley and Uman (1975), Brook and Vonnegut (1960), etc. proved that J-field changes may be positive or negative at higher distances.

Kitagawa and Kobayashi (1959), from all-sky photo multiplier records, have found K-changes by pulses of luminosity. From photographic records, Kitagawa et al. (1962) have occasionally observed the K-changes in a failed leader or in an air discharge just below the cloud base. The K process in which the field change is faster than J process is viewed as a recoil-streamer inside the cloud only, propagating discharge within the cloud. Rakov et al. (1992, 1996), Bils et al. (1988), Villanueva et al. (1994), etc. have shown that the K process is similar to the J process, but has high electric field charge, ramp like structure of field change in between the two strokes and after the last stroke. Kitagawa and Brook (1960) reported that the discharge in K-changes in the cloud discharge and K-changes in the in-cloud portion of the ground discharges are similar. These J and K processes are the initiation of the dart leader, and there is no essential difference between the K-changes that occur during the non-luminous inter stroke periods with M-components (Kitagawa et al., 1962; Mazur et al., 1995; Shao et al., 1995). They also reported that, it is associated either with luminosity inside the cloud, or with a dart leader which propagates downward but fails to reach the ground.

The characteristics of the VHF-UHF radiation associated with K-changes during the inter stroke of CG lightning have been researched by several researchers such as Brook and Kitagawa (1964), Richard et al. (1986), Hayenga (1984), Proctor (1981), Rhodes and Krehbiel (1989), Proctor et al. (1988), Krehbiel et al. (1979). If there is a single stroke but the process of redistribution of charge occurs inside the cloud, it is called F process, or final redistribution process.

References

Adhikari, P. B., Sharma, S. R., & Baral, K. N. (2016). Features of positive ground flashes observed in Kathmandu, Nepal. *Journal of Atmospheric and Solar-Terrestrial Physics, 145*, 106–113.

Adhikari, P. B. (2019). Investigations of the features of electromagnetic fields due to lightning measured in Kathmandu. Ph. D. Thesis, Tribhuvan University.

Anderson, R. B., & Eriksson, A. J. (1980). Lightning parameters for engineering application. *Electra, 69*, 65–102.

Anderson, R. B. (1980). Lightning research in southern Africa. *Transactions on South African Institute of Electrical Engineering, 71*(Part 4), 3–27.

Armstrong, H. R., & Whitehead, E. R. (1968). Field and analytical studies of transmission line shielding. IEEE Transactions, Part III, PAS-87 (pp. 270–281).

Barasch, G. E. (1970). Spectral intensities emitted by lightning discharges. *Journal of Geophysical Research, 75*, 1049–1057.

Beasley, W. H., Uman, M. A., & Rustan, P. L. (1982). Electric fields preceding cloud-to-ground lightning flashes. *Journal of Geophysical Research, 87*, 4883–4902.

Berger, K. (1967). Noval observations on lightning discharges: Results of research on Mount San Salvatore. *Journal of the Franklin Institute, 283*, 478–525.

Berger, K. (1977). *The earth flash, in lightning, physics of lightning*. Academic Press.

Berger, K., & Vogelsanger, E. (1965). Messungen und resultate der blitzforschung der jahre 1955–1963 auf dem Monte San Salvatore. *Bull. Schweiz. Elektrotech. Ver., 56*, 2–22.

Berger, K., & Vogelsanger, E. (1966). Photographische blitzuntersuchungen der jahre 1955–1963 auf dem Monte Sen Salvatore. *Bull. Schweiz. Elektrotech. Ver., 57*, 599–620.

Berger, K., Anderson, R. B., & Kroninger, H. (1975). Parameters of lightning flashes. *Electra, 80*, 223–237.

Bils, J., Thomson, E., Uman, M., & Mackerras, D. (1988). Electric field pulses in close lightning cloud flashes. *Journal of Geophysical Research, 93*, 15933–15940.

Brantley, R.D., Tiller, J. A., & Uman, M. A. (1975). Lightning properties in Florida thunderstorm from vedio tape records. *Journal of Geophysical Research, 80*, 3402–3406.

Brook, M., & Kitagawa, N. (1964). Radiation from lightning discharges in the frequency range 400 to 1000 mc/s. *Journal of Geophysical Research, 69*, 2431–2434.

Brook, M., & Vonnegut, B. (1960). Visual confirmation of the junction process in lightning discharges. *Journal of Geophysical Research, 65*, 1302–1303.

Brook, M., Kitagawa, N., & Workman, E. J. (1962). Quantitative study of strokes and continuing currents in lightning discharges to ground. *Journal of Geophysical Research, 67*, 649–659.

Chen, M., Takagi, N., Watanabe, T., Wang, D., Kawasaki, Z.-I., & Liu, X. (1999). Spatial and temporal properties of optical radiation produced by stepped leaders. *Journal of Geophysical Research, 104*, 27573–27584.

Clarence, N., & Malan, D. (1957). Preliminary discharge processes in lightning flashes to ground. *Quarterly Journal of the Royal Meteorological Society, 83*, 161–172.

Cooray, V., & Lundquist, S. (1985). Characteristics of the radiation fields from lightning in Sri Lanka in the tropics. *Journal of Geophysical Research, 90*, 6099–6109.

Crozier, C. L., Herscovitch, H. N., & Scott, J. W. (1988). Some observations and characteristics of lightning ground discharges in Southern Ontario. *Atmosphere-Ocean, 26*(3), 399–436.

Darveniza, M., & Uman, M. A. (1984). Research into lightning protection of distribution systems II—Results from Florida field work 1978 and 1979. IEEE Transactions on PAS, PAS-103 (pp 673–682).

Davis, S. M. (1999). *Properties of lightning discharges from multiple station wideband electric field measurements.* University of Florida.

Eriksson, A. J. (1978). A discussion on lightning and tall structures. CSIR Special Re-port ELEX 152, National Electrical Engineering Research Institute, Pretoria, South Africa.

Fisher, R. J., Schnetzer, G. H., Thottappillil, R., Rakov, V. A., Uman, M. A., & Goldberg, J. D. (1993). Parameters of triggered-lightning flashes in Florida and Alabama. *Journal of Geophysical Research, 98*, 22887–22902.

Fuquay, D. M., Baughman, R. G., Taylor, A. R., & Hawe, R. G. (1967). Characteristics of seven lightning discharges that caused forest fires. *Journal of Geophysical Research, 72*, 6371–6373.

Garbagnati, E., Giudice, E., & Lo Piparo, G. B. (1978). Measurement of lightning currents in Italy—Results of a statistical evaluation. *ETZ-A, 99*, 664–668.

Golde, R. H. (1945). The frequency of occurrence and the distribution of lightning flash to transmission lines. *AIEE Transactions, 64*, 902–910.

Golde, R. H. (1947). Occurrences of upward streamers in lightning discharges. *Nature, 160*, 395–396.

Golde, R. H. (1967). The lightning conductor. *Journal of Franklin Institute, 283*, 451–477.

Golde, R. H. (1977). The lightning conductor. In: *Lightning, Vol. II: Lightning protection.* Academic Press.

Guo, C., & Krider, E. P. (1982). The optical and radiation field signatures produced by lightning return strokes. *Journal of Geophysical Research, 87*, 8913–8922.

Hagenguth, J. H., & Anderson, J. G. (1952). Lightning to the empire state building. *AIEE Transactions, 71*(3), 641–649.

Hayenga, C. O. (1984). Characteristics of lightning VHF radiation near the time of return strokes. *Journal of Geophysical Research, 89*, 1403–1410.

Hubert, P., & Mouget, G. (1981). Return stroke velocity measurements in two triggered lightning flashes. *Journal of Geophysical Research, 86*, 5253–5261.

Idone, V. P., & Orville, R. E. (1982). Lightning return stroke velocities in the thunderstorm research international program (trip). *Journal of Geophysical Research, 87*, 4903–4915.

Idone, V. P., & Orville, R. E. (1985). Correlated peak relative light intensity and peak current in triggered lightning subsequent return strokes leader characteristics in the thunderstorm research international programme (trip). *Journal of Geophysical Research, 90*, 6159–6164.

Isikawa, H. (1961). Nature of lightning discharges as origins of atmospherics. *Proceedings of Research Institute Atmosphere (nagoya University), 8A*, 1–273.

Jordan, D. M., & Uman, M. A. (1983). Variation in light intensity with height and time from subsequent lightning return stroke. *Journal of Geophysical Research, 88*, 6555–6562.

Jordan, D. M., Idone, V. P., Rakov, V. A., Uman, M. A., Beasley, W. H., & Jurenka, H. (1992). Observed dart leader speed in natural and triggered lightning. *Journal of Geophysical Research, 97*, 9951–9957.

Jordan, D. M., Idone, V. P., Orville, R. E., Rakov, V. A., & Uman, M. A. (1995). Luminosity characteristics of dart leaders and return strokes in natural lightning. *Journal of Geophysical Research, 102*, 22025–22032.

Kitagawa, N. (1957). On the electric field change due to the leader process some of their discharge mechanism. *Papers in Meteorology Geophysics (tokyo), 7*, 400–414.

Kitagawa, N., & Brook, M. (1960). A comparison of intra-cloud and cloud-to-ground lightning discharges. *Journal of Geophysical Research, 65*, 1189–1201.

Kitagawa, N., Brook, M., & Workman, E. J. (1962). Continuing currents in cloud-to-ground lightning discharges. *Journal of Geophysical Research, 67*, 637–647.

Kitagawa, N., & Kobayashi, M. (1959). Field changes and variations of luminosity due to lightning flashes. In: Recent advances in atmospheric electricity, (L. G. Smith edition, pp. 485–501). Pergamon.

Krehbiel, P. R., Brook, M., & McCrory, R. A. (1979). An analysis of the charge structure of lightning discharges to ground. *Journal of Geophysical Research, 84*(C5), 2432–2456.

Krider, E. P., & Guo, C. (1983). The peak electromagnetic power radiated by lightning return strokes. *Journal of Geophysical Research, 88*, 8471–8474.

Krider, E. P., Radda, G. J., & Noggle, R. C. (1975). Regular radiation field pulses produced by intra-cloud lightning discharges. *Journal of Geophysical Research, 80*, 3801–3804.

Krider, E. P., Weidman, C. D., & Noggle, R. C. (1977). The electric field produced by lightning stepped leaders. *Journal of Geophysical Research, 82*, 951–960.

Lin, Y. T., Uman, M. A., Tiller, J. A., Brantley, R. D., Beasley, W. H., Krider, E. P., & Weidman, C. D. (1979). Characterization of lightning return stroke electric and magnetic fields from simultaneous two station measurements. *Journal of Geophysical Research, 84*, 6307–6314.

Livingston, J. M., & Krider, E. P. (1978). Electric fields produced by Florida thunderstorms. *Journal of Geophysical Research, 83*, 385–401.

Lopez, R. E., & Holle, R. L. (1986). Diurnal and spatial variability of lightning activity in Northeastern Colorado and central Florida during the summer. *Monthly Weather Review, 114*, 1288–1312.

Mach, D. M., & Rust, W. D. (1997). Two-dimensional speed and optical risetime estimates for natural and triggered dart leaders. *Journal of Geophysical Research, 102*, 13673–13684.

Mackerras, D. (1978). Prediction of lightning incidence and effects in electrical systems. Electr. Eng. Trans., Inst. Eng. Aust., EE-14, 73–77.

Malan, D. J. (1965). The theory of lightning. In S. C. Coroniti (Ed.), *Problems of atmospheric and space electricity* (pp. 323–331). American Elsevier.

Malan, D. J., & Collens, H. (1937). Progressive lightning iii - the fine structure of return lightning strokes. *Proceedings of the Royal Society of London. Series A – Mathematical and Physical Sciences, 162*, 175–203.

Mazur, V., Krehbiel, P. R., & Shao, X. M. (1995). Correlated high-speed video and radio interferometric observations of a cloud-to-ground lightning flash. *Geophysical Research Letter, 22*, 2613–2616.

McEachron, K. B. (1939). Lightning to the Empire State Building. *Journal of the Franklin Institute, 227*, 149–217.

Muller-Hillebrand, D. (1965). Lightning count measurements in Scandinavia. *Proceeding of IEE (london), 112*, 203–210.

Nag, A., & Rakov, V. A. (2009). Some inferences on the role of lower positive charge region in facilitating different types of lightning. *Geophysical Research Letters, 36*(L05815), 1–05.

Nagai, Y., Kawamata, S., & Edano, Y. (1982). Observation of preceding leader and its downward travelling velocity in Utsunomiya district. *Research Letters on Atmospheric Electricity, 2*, 53–56.

Nakahori, K., Egawa, T., & Mitani, H. (1982). Characteristics of winter lightning currents in Hokuriku district. *IEEE Transactions on Power Apparatus Systems, 101*, 4407–4412.

Orville, R. E. (1968a). A high speed time resolved spectroscopic study of the lightning return stroke: Part I a qualitative analysis. *Journal of Atmospheric Science, 25*, 827–838.

Orville, R. E. (1968b). A high speed time resolved spectroscopic study of the lightning return stroke: Part II a qualitative analysis. *Journal of Atmospheric Science, 25*, 839–851.

Orville, R. E. (1968c). A high speed time resolved spectroscopic study of the lightning return stroke: Part III a qualitative analysis. *Journal of Atmospheric Science, 25*, 852–856.

Orville, R. E. (1968d). Spectrum of the lightning stepped leader. *Journal of Geophysical Research, 73*, 6999–7008.

Orville, R. E. (1975). Spectrum of the lightning dart leader. *Journal of Atmospheric Science, 32,* 1829–1837.

Orville, R. E., & Idone, V. P. (1982). Lightning leader characteristics in the thunderstorm research international programme (trip). *Journal of Geophysical Research, 87,* 11177–11192.

Peckham, D. W., Uman, M. A., & Wilcox, C. E. (1984). Lightning phenomenology in the Tampa bay area. *Journal of Geophysical Research, 89,* 11789–11805.

Prentice, S. A. (1977). Frequency of lightning discharges. *lightning* (Vol. 1, pp. 465–495). Physics of lightning, Academic Press.

Prentice, S. A., & Mackerras, D. (1969). Recording range of lightning-flash counter. *Proceedings of IEE London, 116,* 294–302.

Proctor, D. E. (1971). A hyperbolic system for obtaining vhf radio pictures of lightning. *Journal of Geophysical Research, 76,* 1478–1489.

Proctor, D. E. (1981). VHF radio pictures of cloud flashes. *Journal of Geophysical Research, 86,* 4041–4071.

Proctor, D. E., Uytenbogaardt, R., & Meredith, B. M. (1988). VHF radio pictures of lightning flashes to ground. *Journal of Geophysical Research, 93,* 12638–12727.

Rakov, V. A., & Uman, M. A. (1990). Long continuing current in negative lightning ground flashes. *Journal of Geophysical Research, 95,* 5455–5470.

Rakov, V. A., & Uman, M. A. (2003). *Lightning: Physics and effects.* Cambridge University Press.

Rakov, V. A., Thottappillil, R., & Uman, M. A. (1992). Electric field pulses in k and m changes of lightning ground flashes. *Journal of Geophysical Research, 97,* 9935–9950.

Rakov, V. A., Uman, M. A., Hoffman, G. R., Masters, M. W., & Brook, M. (1996). Bursts of pulses in lightning electromagnetic radiation: Observation and implications for lightning test standards. *IEEE Transactions on Electromagnetic Compatibility, 38,* 156–164.

Rakov, V. A., Uman, M. A., Rambo, K. J., Fernandetz, M. I., J., F. R., Schnetzer, G. H., & Bondiou-Clergerie, A. (1998). New insights into lightning processes gained from triggered-lightning experiments in Florida and Alabama. *Journal of Geophysical Research, 103,* 14117–14130.

Rakov, V. A., Crawford, D. E., Rambo, K. J., Schnetzer, G. H., Uman, M. A., & Thottappillil, R. (2001). M-Component mode of charge transfer to groud in lightning discharges. *Journal of Geophysical Research, 106,* 22817–22831.

Rhodes, C., & Krehbiel, P. R. (1989). Interferometric observations of a single stroke cloud to ground flash. *Geophysical Research Letter, 16,* 1169–1172.

Richard, P., Delannoy, A., Labaune, G., & Laroche, P. (1986). Results of spatial and temporal characterization of the VHF-UHF radiation of lightning. *Journal of Geophysical Research, 91,* 1248–1260.

Schonland, B. F. J., & Collens, H. (1934). *Progressive Lightning. Proc. Roy. Soc., A143,* 654–674.

Schonland, B. F. J., Malan, D. J., & Collens, H. (1935). Progressive lightning ii. *Proceedings of Royal Society A, 152,* 595–625.

Schonland, B. F. J. (1956). The lightning discharge. In *Handbuck der Physik,* (Vol. 22, pp. 576–628). Springer-Verlag.

Shao, X. M., Krehbiel, P. R., Thomas, R. J., & Rison, W. (1995). Radio interferometric observations of cloud-to-ground lightning phenomena in Florida. *Geophysical Research Letter, 100,* 2749–2783.

Shao, X. M. (1993). The development and structure of lightning discharges observed by vhf radio interferometer. New Mexico Institute of Minerals and Technology, Socorro.

Shindo, T., & Uman, M. A. (1989). Continuing current in negative cloud-to-ground lightning. *Journal of Geophysical Research, 94,* 5189–5198.

Thomson, E. M. (1985). A theoretical study of electrostatic field wave shaves from lightning leaders. *Journal of Geophysical Research, 90,* 8125–8135.

Thottappillil, R., Rakov, V. A., & Uman, M. A. (1990). K and m changes in close lightning ground flashes in Florida. *Journal of Geophysical Research, 95,* 18631–18640.

Thottappillil, R., Rakov, V. A., Uman, M. A., Beasley, W. H., Master, M. J., & Shelukhin, D. V. (1992). Lightning subsequent stroke electric field peak greater than the first stroke peak and multiple ground terminations channel base. *Journal of Geophysical Research, 97*, 7503–7509.

Thottappillil, R., Goldberg, J. D., Rakov, V. A., Uman, M. A., Fisher, R. J., & Schnetzer, G. H. (1995). Properties of m components from currents from currents measured at triggered lightning channel base. *Journal of Geophysical Research, 100*, 25711–25720.

Thottappillil, R., Schoene, J., & Uman, M. (2001). Return stroke transmission line model for stroke speed near and equal that of light. *Geophysical Research Letters, 28*, 3593–3596.

Uman, M. A., & McLain, D. K. (1969). Magnetic field of the lightning return stroke. *Journal of Geophysical Research, 74*, 6899–6910.

Uman, M. A. (2001). The lightning discharge (p. 377). Dover Publications.

Villanueva, Y., Rakov, V., & Uman, M. (1994). Microsecond-scale electric field pulses in cloud lightning discharges. *Journal of Geophysical Research, 99*, 14353–14360.

Wang, D., Rakov, V. A., Uman, M. A., Takagi, N., Watanabe, T., Crawford, D. E., & Kawasaki, Z. I. (1999). Attachment process in rocket-triggered lightning strokes. *Journal of Geophysical Research, 104*, 2143–2150.

Weidman, C. D., & Krider, E. P. (1979). The radiation field wave forms produced by intra-cloud lightning discharge processes. *Journal of Geophysical Research, 84*, 3159–3164.

Chapter 3
Lightning Types

3.1 Introduction

Lightning is the most spectacular and fascinating atmospheric event. Lightning phenomenon is believed to prevail in earth much before life began on the planet. Rakov and Uman (2003) reported that life evolved on our planet more than 3 billion years ago. For the development of life, the molecules of hydrogen cyanide (HCN) were required in earth's atmosphere. Lightning produced the molecules such as hydrogen cyanide (HCN) in the atmosphere. However, the scientific study of lightning started only from the latter half of the eighteenth century. The systematic study of lightning began in 1752, with two separate experiments, known as sentry-box experiment and kite experiment, performed by Benjamin Franklin which proved that the lightning is an electrical discharge phenomenon. All ancient human civilizations have explained lightning according to their religious beliefs. They described various myths and pranks about lightning in different civilization (MacGorman & Rust, 1998; Rakov & Uman, 2003; Uman, 2001). Lightning has played a prominent part in almost all ancient religions (Andrews et al., 1992). Since lightning phenomenon is very difficult to study, Andrews et al. (1992) explained that it has a great deal of mystery and mythology. Ogawa (1995) stated that the fascinating display due to lightning has immeasurable effects on all the things that exist; not only near the lightning but also far from it. Lightning can cause severe damages to physical structures, human life, animals, transmission lines, transmission towers, communication towers, and other tall physical structures on the ground, which are more vulnerable to the lightning activities. It is estimated that over 2000 lightning discharges take place at any given time around the globe, and more than 2000 people lose their lives every year globally.

P. B. Adhikari and A. Adhikari, *Lightning Discharges*,
SpringerBriefs in Applied Sciences and Technology,
https://doi.org/10.1007/978-981-19-1926-8_3

3.2 Types of Lightning

A lightning discharge in the atmosphere has a long channel with many branches extended vertically as well as horizontally. As already mentioned in 2.3, the charge structure in the cloud is believed to be tripolar type, in which the main negative charge remains at the center of the cloud, equal amount of positive charge remains at the top of the cloud and a small positive pocket charge develops at the bottom of the thunder-cloud. Whether lightning strikes on ground or not, it is commonly called a lightning flash. If the lightning flashes reach from cloud to the ground or strikes the object of the ground—tall trees, tall buildings, high towers, etc., they are called lightning strikes. The lightning strikes include the stepped leader moving downward and the return stroke in upward direction. Besides this, the lightning discharges within the same cloud or on another cloud or just outside on the air are called cloud discharges or cloud-to-cloud lightning or intra-cloud discharges (ICs). Hence, lightning discharge is initiated from a thunder cloud and can be divided into two categories, cloud-to-ground (CG) discharges and cloud discharges. The lightning discharges, which terminate at the ground, are called cloud-to-ground discharges and the discharges, which do not reach the ground, are in general called the cloud discharges (Berger, 1977). According to Berger (1977), the downward negative lightning transports nega-tive charges from the main negative charge center to ground and accounts for 90% of ground flashes. Less than the 10% of ground flashes are downward positive lightning which transports positive charges to ground from the main positive charge center. The remaining two upward lightnings are very rare, as opposed to the downward light-ning, which occur due to the presence of tall structural objects of height more than one hundred and fifty meters. In the tripole structure of the cloud shown in Fig. 3.1,

Fig. 3.1 Tripole structure of the cloud, **a** represents intra-cloud lightning, **b** represents negative cloud to ground lightning, **c** represents positive cloud to ground lightning, and **d** represents cloud to air discharge

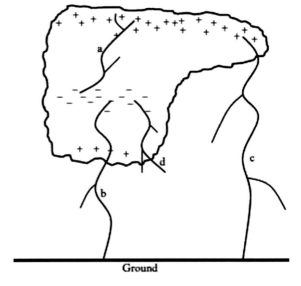

"a" represents intra-cloud lightning, "b" represents negative cloud to ground lightning, "c" represents positive cloud to ground lightning, and "d" represents cloud to air discharge.

3.3 Cloud Discharges (ICS)

Among the lightning discharges, the cloud discharges account for about three-quarters of the total lightning. The cloud discharge is the discharge in which lightning phenomena occur without connecting to the earth's surface. It comprises three types of lightning: intra-cloud, inter-cloud, and cloud to air discharges. Among these three types, the discharge that occurs in the positive and negative charge within the same cloud frequently are called intra-cloud discharges. The inter-cloud discharge occurs between the positive charge of thunder cloud and the negative charge of another thunder cloud. The third, cloud to air discharge occurs between the thunder cloud and clear air outside the cloud.

Overall, the process electric field change on intra-cloud and cloud to air is similar (Ogawa & Brook, 1964). For each of the lightnings, either cloud to ground discharge or cloud to cloud discharge, the initial process of electrical discharges are same. Williams (1989) reported that about more than ten cloud flashes occur before the occurrence of CG lightning. Although the cloud flashes occur more frequently, they have much less attention to the research and hence they are less understood, mostly because the cloud discharge does not pose any threat of death or injury to human beings, animals, and living things on the ground. They can't reach the ground, and it is challenging to record their photograph; only the optical observation gives useful information about it (Isikawa, 1961; Mackerras, 1968; Ogawa & Brook, 1964; Takagi, 1961). It is difficult to measure the charge transfer and the currents in the cloud discharges. It should not be underestimated because of the threat to airplanes, avionics, and electromagnetic devices. The study of cloud flashes is also equally important as the ground flashes to understand the physics of lightning discharge. Some of the researchers, Kitagawa and Kobayashi (1959), Kobayashi et al. (1958), Takagi and Takeuti (1963), etc. have measured both the electric field pulses and the emission of light; Krehbiel et al. (1979), Liu and Krehbiel (1985), etc. used the multiple-station electric field network to determine the accurate location of the lightning. Proctor (1971, 1981, 1997), Hayenga (1984), Richard and Auffray (1985), Mazur (1989), Shao and Krehbiel (1996), Rison et al. (1999), Thomas et al. (2000) etc. used VHF and UHF lightning locating system and obtained the image of time-resolved intra-cloud channels. Similarly, the electric field measurement, acoustic measurement, and RADAR observation were also used by Weber et al. (1982), Mazur et al. (1984), etc. for more information about the cloud flash.

Rakov and Uman (2003) explained the cloud flash as an electrode-less discharge phenomenon by using optical observation. As we know the tripole structure of the cloud, the cloud discharge occurs mostly from the lower and upper boundaries of a negative charge center, and it bridges the two charge centers—the upper positive

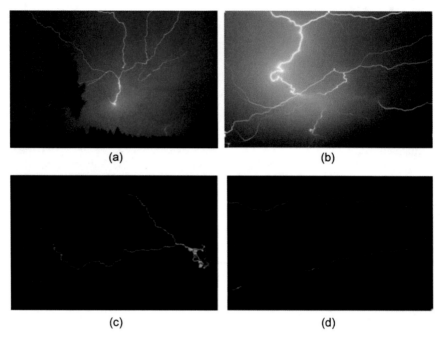

Fig. 3.2 **a–d** All are the examples of cloud flashes showing variability from flash to flash

charge, and the main negative charge. The cloud flashes vary from flash to flash and storm to storm. Some examples of such cloud flashes are represented in Fig. 3.2 (Adhikari & Sharma, 2020).

Initially, Kitagawa and Brook (1960) classified the cloud flashes into three stages, according to the occurrence in a flash as "initial," in the middle as "very active" and the last as "final" stage. However, Bils et al. (1988) and Villanueva et al. (1994) explained that there are only two stages: the initial stage, which is the most active part of the cloud, and the final stage. They also suggested that from the Kitagawa and Brook (1960) data, cloud flashes have two stages only. They supported the theory that the active stage with the initial breakdown stage has large slow field variation with large microsecond scale bipolar pulses. The examples of one stage, two-stage, and the expansion of first and second stages of two-stage flashes are shown in Fig. 3.3 (Adhikari & Sharma, 2020). The fine structures of negative initial polarity with large IB pulses train and expansion of final stage of flash containing fine pulses with both polarities are shown in (c) and (d) respectively in Fig. 3.3.

Adhikari and Sharma (2020) have reported that the three-stage flashes are very rare, only about 3%. However, these cloud flashes are continuous without any stages, which is shown in the first diagram (a), and the expansion of so-called the first, second, and third stages are shown in (b), (c), and (d) in Figure 3.4.

Rakov and Uman (2003) reported that the early stage involves the channel of negative charge moving in the order of 10^5 m/s. The stepped leader process in negative

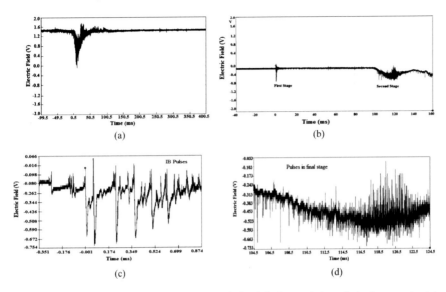

Fig. 3.3 Examples of electric field signature of cloud flash consisting of single stage in (**a**), consisting of two stages in a single flash is shown in (**b**), and the examples of expansion of electric field signature of first and second stages of two stages flashes are shown in (**c**) and (**d**) respectively

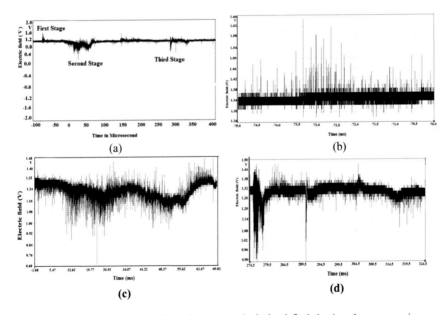

Fig. 3.4 An example of electric field due to a typical cloud flash having three stages in **a** an expansion of first stage, second stage and third stage of three-stage typical cloud flash of (**a**) shown in (**b**), (**c**), and (**d**)

CG flashes is quite similar to the early stage process in the initial breakdown process. For the cloud flash, the upper and lower boundaries of the negative charge center are important. Then, the upper boundary of the negative charge is connected to the upper positive charges and forms the lightning of positive electric field change. That means the discharge between the upper boundary of the main negative charge and the upper positive charge region can be regarded as higher-origin cloud flashes. Similarly, the lower boundary of negative charge is connected with the lower positive charges pocket and forms the lightning of negative electric field change in the cloud flash called lower origin flashes. Shao and Krehbial (1996), Rison et al. (1999), Rakov and Uman (2003), Uman (1987), Proctor (1981, 1997), Gomes et al. (2004), etc. reported that the cloud discharges occur in a single vertical channel which bridges the two regions of the charges. The height of the origin of the cloud flash phenomena gives the characteristics of the stepped leaders process and also the initial breakdown process. The cloud discharge begins, from the two boundaries, with bidirectional leaders. Rakov and Uman (2003) described that the upper positive and negative center charges in the cloud are called main charges, with a magnitude of 40 Coulombs on each. Only 3–5 Coulombs of positive charge remains at the bottom, as a pocket charge as shown in Fig. 3.5.

In the cloud discharge, generally, there are two stages. As already mentioned, the second or final, or late stage is also called the J-type stage. The process in the late stage of the cloud discharge is similar to the junction process (J process) in the cloud to ground discharges. This channel is generally positively charged, starting from low speed in the order of 10^4 m/s (Rakov & Uman, 2003).

Shao and Krehbiel (1996), Rison et al. (1999) reported that most of the radiation sources in the cloud discharges were located at the upper level. However, the cloud flashes can take place between the main negative charge center and the lower positive charge pocket. Proctor (1981, 1997) and some other scholars reported that the pulses cloud discharge in higher-origin flashes is radiated at a lower rate of very high frequency and ultra-high frequency compared to lower-origin flashes. However, the

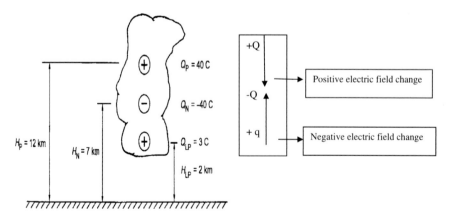

Fig. 3.5 An idealized vertical tripole structure of a thunderstorm (Rakov & Uman, 2003)

lower-origin cloud flashes are indistinguishable from the stepped leaders in the cloud to ground flashes (Proctor, 1997). The initial polarity, separation of pulses, pulse width, and pulse amplitude of chaotic pulse trains (CPTs) were studied by Gomes et al. (2004). According to Gomes et al. (2004), there is a high probability of 2–4 μs of pulse width between 1 and 10 μs and pulse separation of 2–20 μs. The intra-cloud lightning waveforms due to the radiation field are alike within the cloud but differ from the return stroke of cloud to ground lightning (Weidman & Krider, 1979). Krider et al. (1975) reported that the FWHM of the waveform of the radiation field is 0.75 μs, and the time interval between successive pulses is 5 μs. However, the time interval between the successive pulses for the 50 ms pulse train is to be 147 μs, as reported by Gomes and Cooray (2001). They further reported the two pulses: single peaked pulses with a pronounced positive overshoot and multiplied peaked pulses with less pronounced overshoot. The sequences of single polar pulses produced by dart-stepped leaders are reported by Krider et al. (1975); while the bipolar pulses with structures on their fronts reported by Weidman and Krider (1979); and the bipolar pulses without such structures reported by Le Vine (1980) and Cooray and Lundquist, (1985). The overall characteristics of cloud flashes are represented in the following Tables 3.1 and 3.2.

The characteristics of preliminary breakdown pulses in the negative cloud to ground discharge were described by Nag and Rakov (2008a). That is the initiation of downward negative stepped leaders but fails to propagate to the ground. The excessive amount of positive pocket charge in the tripole structure may convert the cloud to ground flash to an intra-cloud one (Nag & Rakov 2009). Based on preliminary breakdown pulse trains, the multiplicity of the strokes, leader propagation mode, branching of the in-cloud channel, and occurrence of continuing currents, Nag and Rakov (2012) found many differences between the positive and negative flashes.

Bitzer et al., (2013) reported about the waveforms due to radiation sources from lightning which are used to determine the space–time position. They compared the National Lightning Detection Network (NLDN) and the North Alabama Lightning Mapping Array (NALMA). Karunarathne et al. (2013) observed the initial break-down pulses (IBPs) in the fast electric field change at the beginning of intra-cloud (IC) and cloud-to-ground (CG) lightning flashes. The preliminary breakdown pulse characteristics of cloud to cloud lightning flashes of the summer thunderstorm in Japan were reported by Wu et al., (2015). He also reported that the initiation alti-tudes ranges of preliminary breakdown pulses in intra-cloud lightning flashes range from 5 to 10 km, and preliminary breakdown pulses in negative CG flashes range from 4 to 7 km. IC flash occurs between the main upper positive and main negative charges and propagates upward with a speed of initial leader $(0.5$ to $17.8) \times 10^5$ m/s with an average of 4×10^5 m/s. So, the altitude should be considered when discussing pulse characteristics in IC flashes (Wu et al., 2015). Wu et al. (2019) researched a special type of IC flash initiated from a high altitude above 12 km. The downward positive intra-cloud flashes generally initiated from about -500 °C moving with the speed of 10^4 m/s dominate the upward negative leader's IC flashes. Adhikari and Sharma (2020) have explained that the investigation of the characteristic features of electric field radiated by the cloud flashes. The initial polarity pulses that lower the

Table 3.1 Overall characteristics of cloud flashes

Researcher	Place	Duration of flash (ms)	Total pulse duration (μs)	No. of flashes	Total width (μs)	Half peak width (μs)
Pierce (1955)	England	245		685		
Takagi (1961)	Japan	300				
Isikawa (1961)	Japan	420	40	616		
Ogawa and Brook (1964)	New Mexico	500				
Mackerras (1968)	Australia	480				
Weidman and Krider (1979)			63	137		
Le Vine (1980)	Florida		10 - 20		10	
Cooray and Lundquist (1985)	Sri Lanka		75	26	13	
Bils et al. (1988)	Florida	660	74	89	–	9.3
Willett et al. (1989)	Florida			24	10–15	2.4
	Florida		> 13	156 (-ve)	> 4.7	1.8
			> 22	10 (+ve)	> 7.7	1.6
Villanueva et al. (1994)	Florida		53	17	25	2.8
			61	6	27	3.3

Table 3.2 Some parameters for the cloud activity

Parameters	I B Pulse	Cloud activity for		Overshoot
		Positive initial	Negative initial	
Rise time	0.93 μs	–	–	6.29 μs
Total duration	1.9 μs	168.3 ms	231.1 ms	146.63 ms
Interpulse interval	167.41 μs	309.79 μs	211.42 μs	
Amplitude	131.03 mV			33.76 mV
Time of activity for first stage = 11.23 ms				
Time of activity for second stage = 66.79 ms				

Table 3.3 A comparison of the characteristics of cloud flashes (Adhikari & Sharma, 2020)

Researcher	Inter-pulse interval (Average, μs)	Time for first stage	Time of cloud flash
Gomes and Cooray (2001)	147 μs	–	–
Sharma et al. (2005)	35.7 μs	10 ms	100–400 ms
Weidman and Krider (1979)	780 μs for negative 130 μs for positive	–	–
Adhikari and Sharma (2020)	309.79 μs for positive 211.42 μs for negative	11.23 ms	80–469.5 ms

negative charge to the ground are termed negative polarity pulses, and the pulses that lower the positive charge to the ground are termed positive polarity pulses.

Adhikari and Sharma (2020) have reported that the total duration of the cloud flash, in 500 ms time window, varied from 80 to 469.5 ms and on comparing with the cloud flashes from the temperate Sweden thunderstorm, varied from 100 to 400 ms that gave similar result. The Himalayan country and temperate country have found quite similar results, as mentioned in Table 3.3. The average duration of the first initial activity is 11.23 ms, also comparable with 10 ms in Sweden. Similarly, the duration of the second stage of cloud activity, the gap between the first and second stage, the average time interval between successive pulses are also compared with others (Adhikari & Sharma, 2020).

3.4 Cloud to Ground Discharges (CGs)

The transport of either positive or negative charge of the cloud to the ground is called cloud to ground discharge (CG). It is believed that among all the lightnings, only 25–30% are cloud to ground lightning discharges. Berger (1977) categorized cloud to ground lightning into four different parts according to the charge transports from cloud to ground and the direction of the leaders moving upward or downward. They are:

1. Downward negative lightning
2. Downward positive lightning
3. Upward negative lightning
4. Upward positive lightning.

Figure 3.6 illustrates these four different types of lightning in which the first downward negative lightning is believed to occur in about 90% of the total cloud to ground discharges. In this discharge, the negative discharge of the cloud transports toward the ground, and the stepped leader also moves downward, as shown in Figure (a). Similarly, when the positive charge of the cloud (about 10% of the CGs lightning)

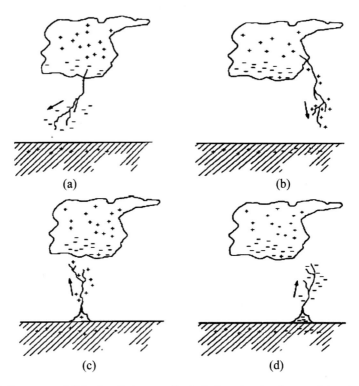

Fig. 3.6 Four types of lightning discharges effectively lowering the charges of the cloud to the ground. **a** downward negative lightning discharges, **b** downward positive lightning discharges, **c** upward negative lightning discharges, and **d** upward positive lightning discharges

transports to the ground, and the stepped leader also starts from cloud to ground, which is shown in Figure (b). The third and fourth types of CGs lightning discharges are the charges that transport from cloud to the ground, but the stepped leaders are moving from ground to cloud. They are rare, but they occur on high mountains, hills, high towers, and tall trees having a height of about 150 m and more. These are illustrated in Figures (c) and (d) in Fig. 3.6.

3.4.1 Downward Negative Lightning Discharges

The negative charge transfer to the ground in negative CG lightning is researched frequently. They are best understood as their current produces most of the damages, and it is optically the brightest, producing a large and easily identifiable electro-magnetic signature (Uman, 2001). Pierce (1955) found that almost all the recorded field changes can be most simply explained in terms of leaders carrying a negative charge and originating in the lower part of thundercloud. Those reaching the ground are succeeded by return strokes. It is considered that both leaders and return strokes consist of a central core surrounded by an envelope of corona. About 99% of the ground flashes are estimated to be cloud-to-ground type and exclusively strike the objects on the ground with height of several tens of meters. So, the current measurements from high towers do not truly represent the majority of the cloud-to-ground lightning current. In order to get the knowledge about current due to the majority of return strokes and other processes, an indirect measurement could be used, which involves the remote sensing of the signatures emitted from the channel. In fact, the remote electric and magnetic field parameters are used to deduce the current. Sharma (2007) explained that with the advancement of the measuring equipment, especially their higher resolution and fast responses with larger bandwidths, many new events have been unveiled and consequently further challenges.

Generally, the lower positive charge region enhances the electric field at the bottom of the main negative charge region and facilitates the launching of the negative charge toward the ground. However, Nag and Rakov (2009) explained that the presence of excess lower positive charge region might prevent the occurrence of negative cloud to ground discharge. Nag et al. (2009) examined microsecond and submicrosecond scale pulses. The majority of pulses in both cloud and ground discharges analyzed in the study were associated with the initial breakdown process and relatively small in amplitude and duration when compared with negative cloud to ground lightning discharges and pulse ground associated with cloud to ground leaders. Nag and Rakov (2009) determined that the pulse train in ground discharge contains a larger fraction of narrow pulses than the pulse train in the attempted cloud to ground leaders. According to Williams (1989), the lightning activity itself follows a specific pattern with the intra-cloud (IC) lightning normally appearing in the developing stage followed by cloud-to-ground (CG) lightning during the mature stage; however, both types of lightning have the possibility of occurrence in the decaying stage of thunderstorms. Price (2008) has stated that lightning in thunderstorms is strongly linked to the microphysics and dynamics of thunderstorms, and hence, charges in the lightning activity can tell us about changes in the internal processes within the thunderstorms.

Most researchers have reported the downward negative lightning discharges, which occur about 90% of the time or more. Starting from Benjamin Franklin in 1752, who proved that lightning is a discharge phenomenon, many scientists have researched it since then. After 140 years only, Hoffert in England in 1889, Weber and Walter in 1902 in Germany, used the photography method to analyze the lightning phenomena. At the same time, Pockels 1897, in Germany, used the second process of current measurement of lightning to study it. Then, Wilson 1916, in England, who won a Nobel Prize, used the electric field measurement to estimate the charge structure in the cloud. As already mentioned, most of the lightning from cloud to the ground is downward negative lightning. The overall discharge of the cloud to ground lightning is called flash, and it is composed of several processes, as mentioned in the previous chapter. Among these processes, the strokes are the most important one, and each flash contains about three to five strokes generally on average, but until now, up to 26 strokes have also been found in a single flash (Rakov & Uman, 2003). Each flash of downward negative lightning contains a stepped leader and return stroke.

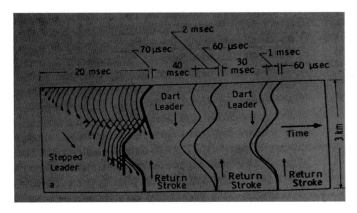

Fig. 3.7 A drawing of the luminous features of lightning flashes below a 3-km cloud base adapted from Uman (2001)

However, it may occur as a stepped leader, return stroke, dart leader, continuing current, M process, J process, K process, etc. The example of the entire process of lightning in the order with the time taken is also represented in Fig. 3.7.

Bazelyan and Raizer (2000) reported that the potential difference between the cloud charge source and ground is in the order of some tens of megavolts and the electric potential difference between the tips of the downward moving stepped leader and ground are also comparable, i.e., about some tens of megavolts. As we know, the line integral of the electric field intensity gives the magnitude of potential difference between the cloud charge and the ground. As the leader approaches the ground, the distance decreases, and hence the electric field intensity increases. When it exceeds the critical value, then the attachment process occurs. Probably some tens of meters above the ground, the downward moving stepped leader, and the upward moving response leader meet at a point. The attachment process ends when they come in contact with each other; then the return stroke process starts, in which the neutralization process occurs. During this time between the attachment process and return stroke process in the lightning, it has complex phenomena. The initial process of starting the phenomena of lightning is called initial breakdown or preliminary breakdown. This process initiates the downward moving stepped leader in the cloud. It is the unique lightning process inside the cloud which occurs just before the emergence of the stepped leader from the base of the cloud, and it is observed by the luminosity produced for hundred or more milliseconds. During this time, the electric field charge has a large duration, more than 100ms before the main stroke. The initial breakdown process is the vertical discharge phenomena between the main negative charge and the lower pocket positive charge stated by Clarence and Malan (1957).

The initial breakdown process occurs as a succession of breakdown events prior to the development of leader to ground (Krehbiel et al., 1979). They used eight station electric field measurements and found two flashes in New Mexico, which transported the negative charge of 2 °C, and 3.6 °C moved over 1.5 km and 3.8 km with average currents 44 A and 70 A, respectively. Rhodes (1989) explained the breakdown stage as the initial steps, the intermediate stage which is the initiation of the stepped leader and the stepped leader just before the return stroke using different techniques. The initial breakdown process may depend on the type of storm, winter and summer stage of the storm's life cycle, the distance of the cloud, the latitude of the places, propagation effects, etc. (Brook, 1992). Rakov and Uman (2003) reported that the rise time of the individual pulses is 10 μs on average, and the total duration is 20–40 μs. However, the rise time of the individual pulse in the Himalayan region is only 2.6 μs, and the total duration of the pulse is about 15 μs (Adhikari & Adhikari, 2021). Weidman et al., (1981); Willett et al. (1990), etc. studied the preliminary breakdown stage using the frequency spectra, optical radiations, and different frequency radiations.

The leader steps have some properties. The longer pause times should be followed by longer step lengths; for example, pause time lies between 29 and 52 μs for the step lengths between 3 and 50m (Berger & Vogelsanger, 1966). If there are more branches in the process of stepping downward in the leader, then the time between the pulses decreases as the stepped leader approaches the ground. The time between the steps is 50 μs far above the ground, but as the leader approaches the ground, the time decreases to 13 μs (Kitagawa, 1957). Using remote magnetic field measurements, the average stepped leader currents measured the two leaders and were found to be 50 and 63 A (Williams & Brook, 1963). Similarly, Thomson (1985) measured the current of 62 stepped leaders from the electric field measurement in Florida and found an average of 1.3 kA with the range of 100 A to 5 kA. Similarly, Krehbiel (1981) also found the current for 7 stepped leaders to be 1.3 kA with the range of 200 A to 3.3 kA in Florida by multiple-station electric field measurements.

In this stepped leader process, the downward moving stepped leader and the upward moving response leader initiated from the ground or grounded object are attached. It may be possible that there are two or more initiated upward leaders moving upward from the grounded object or ground. Also there is a possibility of downward moving stepped leader touching or attaching with anyone from the upward initiated leader. This upward initiated leader is called upward connecting leader. But those initiated upward leaders which are not connected to the downward moving stepped leader are called unconnected upward leaders or unconnected streamers (Rakov, 1998). The attachment process occurs for the first stroke in the virgin air, but the subsequent strokes occur in the previous lightning channel. There are two processes, time-resolved optical image and still photograph method, to observe the attachment process. Yokoyama et al. (1990) used the time resolve optical image process to study the attachment process. Golde (1947) studied the attachment of first and subsequent strokes by using the time-resolved optical image approach. However, later he used the still photograph to study the attachment process in the lightning flash.

In the attachment process, the upward leaders and downward leaders make contact and form the plasma channel. These high conducting plasma channels move toward each other. Rakov (1998) explained that one plasma channel wave moves toward the ground in the downward direction and another channel wave moves in an upward direction through the leader's path. The downward-directed return stroke wave, due to the short distance, reaches quickly to the ground and reflects forming the return stroke wave, and the combined wave reaches the cloud. This hot plasma channel upward is reflecting return stroke wave. This wave moves faster due to the original path, meets the previous i.e., the return stroke, is the main function of the lightning process (Rakov, 1998). Hence, in the short period, there are two hot channel return stroke waves moving in two directions, also called bidirectional leader, in the attachment process. It is the most important phenomenon of the lightning flash, which is visible due to the brightest lightning process outside the cloud. The phenomena produce the lightning electromagnetic signature, which is very important to study. So, there are many researchers in the world involved in studying it. The properties of negative cloud-to-ground lightning flash, properties of stepped leader, dart leader and stepped dart leader, properties of return stroke, subsequent stroke, continuing current and M-component and return stroke channel with physical properties such as temperature, pressure, electron density, etc. are illustrated in the following Table 3.4.

3.4.2   Downward Positive Lightning Discharges

The lightning flash that transports the positive charge from cloud to ground is called positive ground flash. It occurs very rarely; about 90% are believed to be negative cloud to ground lightning, and less than 10% of the total cloud to ground lightning flashes are downward positive lightning. However, the other upward positive lightning discharges and upward negative lightning discharges are even rarer. The tripole charge structure of a thundercloud is explained in Section 2.3; however, it still remains a mystery. Rakov and Uman (2003) described that positive CG lightning depends on the variability of the latitude, topography, season, and type of storm. One of the examples of the lightning data of monthly and annual percentage of positive lightning flashes for six years is given in Table 3.5, which indicates that positive lightning occurs with a high percentage in winter seasons. Similarly, positive lightning flashes depend on the variability of the latitude, and it has been reported from various locations by Mackerras and Darveniza (1994), which is represented in Table 3.6. The lightning discharge channel with many branches acts as a transmitting antenna for electromagnetic waves of different ranges of frequencies (Ogawa, 1995).

Nag and Rakov (2012) described that the positive ground flashes are considerably less studied and, hence, less understood than the negative ground flashes. As already mentioned, the positive lightning ground flashes are about 10% of the total ground flashes. However, Baral and Mackerras (1993) found that 28% of the total ground flashes are positive ground flashes along the Himalayas range. The positive ground flashes depend on different geographical locations, different latitudes, and different seasons. In the Himalayan regions, Baral and Mackerras (1993) found that 28% of the total ground flashes are the average positive CGs with the maximum occurrence of 38% positive lightning, which was in the post-monsoon season. The theory of lightning requires understanding the physical processes in the electrical breakdown of the air, BIL structure as a microscale process, and the propagation of

Table 3.4 Features of negative cloud-to-ground lightning and properties of return stroke

(a) Properties of stepped leader, dart leader and stepped dart leader

Parameters	Stepped leader	Dart leader	Dart-stepped leader
Step length (m)	50	–	10
Time interval (μs)	20–50	–	5–10
Step current (kA)	>1	–	–
Step charge (mC)	>1	–	–
Average propagation speed (m/s)	2×10^5	$(1–2) \times 10^7$	$(1–2) \times 10^6$
Total duration (ms)	35	1–2	–
Average current (A)	100–200	1	–
Total charge (C)	5	1	–
Electric potential (MV)	~50	~15	–
Channel temperature (K)	~10,000	~20,000	–

(b) Properties of return stroke, subsequent stroke, continuing current and M-component

Parameters	Return stroke	Subsequent RS	Continuing current	M-component
Peak current	30 kA	10–15 kA	100–200 A	100–200 A
Max. current rate (kA/μs)	> 10–20	100	–	–
Rise time of current (μs)	5	0.3–0.6	–	300–500
Current duration to half peak (μs)	70–80 μs	30–40 μs	100 ms	–
Charge transfer (C)	5	1	10–20	0.1–0.2
Propagation speed (m/s)	$(1–2) \times 10^8$	$(1–2) \times 10^8$	–	–
Channel radius (Cm)	(1–2)	1–2	–	–
Channel temperature (K)	~30,000	~30,000	–	–

(c) Properties of negative cloud-to-ground lightning flash

Parameter	Overall flash
Duration, (ms)	200–300
Number of strokes per flash	3–5
Inter-stroke interval. (ms)	60
Charge transfer (C)	20
Energy, (J)	$10^9–10^{10}$

(d) Physical properties during return strokes, before and after return stroke

Channel characteristics	Before RS	During RS	After RS
Temperature	20,000 K	> 30, 000 K or 31,000–36,000 K	3000 K
Electron density	8×10^{17} cm^{-3} (within 5 μs)	$(1–1.5) \times 10^{17}$ cm^{-3} at 25 μs	1×10^{17} cm^{-3} at 50 μs

(continued)

Table 3.4 (continued)

(d) Physical properties during return strokes, before and after return stroke

Channel characteristics	Before RS	During RS	After RS
Pressure	8 atm	1 atm	1 atm

Table 3.5 Month and annual percentage of positive lightning flashes

	Percentage of positive lightning flashes					
Month	1992	1993	1994	1995	1996	1997
January	13.4	10.3	15.6	20.2	24.5	19.8
February	11.9	10.3	12.6	14.3	19.5	22.7
March	9.1	8.7	11.3	14.9	15.8	19.2
April	6.6	9.4	7.8	17.2	15.9	15.4
May	4.6	6.0	6.3	14.7	14.3	12.6
June	4.1	5.3	4.0	8.2	8.6	11.1
July	3.5	3.7	4.0	6.5	8.0	8.1
August	3.0	3.2	3.7	6.6	7.3	6.9
September	3.9	3.6	4.6	7.0	10.6	7.0
October	5.4	5.9	8.7	12.1	15.1	14.1
November	9.4	12.6	15.2	18.4	17.6	15.9
December	21.5	24.9	16.5	19.3	17.5	19.3
Annual average	4.2	4.6	4.9	9.3	10.2	10.1

the flash macroscale process. The flash can cover the path of hundreds of kilometers (MacGorman & Rust, 1998). Bazelyan and Raizer (2000) raised some problems about the initiation of discharge process in the lightning phenomena, charge distribution in the cloud, initial breakdown voltage, state of the gas in the lightning channel, etc., which are unsolved until now.

When the cloud charge is horizontally separated, and the positive charge and the negative charge are far away, then positive cloud to ground lightning flashes may be initiated from the upper positive charge in the thundercloud (Uman, 2001). The primary source of lightning threat is the current due to return stroke and its associated parameters. Rakov and Uman (2003) reported the measurements of lightning current carried out from the high television towers in Italy and Salvatore, Switzerland.

Using the electric field measurement system, Rust et al. (1981) explained that the positive cloud to ground lightning flashes generally come from the top of the cloud. Most flashes contain single stroke and continuing currents following the ground stroke. By using the combination of electric field and optical measurements, Fuquay (1982) reported that all the positive lightning flashes were composed of a single stroke and initiated by downward moving positive leaders. Cooray (1984) compared the electric field measurements of a negative and positive lightning flash and found a

Table 3.6 Cloud flash percentage (F_C), positive ground flash percentage (F_{+ive}), and mean annual total flash density (V_F) as a function of latitude. (Adhikari, 2019a)

Location	Latitude	Period (year)	F_c (%)	F_{+ive} (%)	V_F (Flashes, km^{-2} y^{-1})
Singapore	1.2 °N	4.5	77	2	61
Bogota, Columbia	5.0 °N	1.0	33	2	19
Lae, PNG	6.4 °S	0.6	57	3	–
Darwin, Australia	12.2 °S	3.0	58	2	10
Brazil	23.2				
Stanwell, Australia	23.5 °S	4.0	62	2	5.6
Gaborone, South Africa	24.4 °S	1.1	52	4	16
Ackland, Australia	27.3 °S	0.3	79	9	–
Brisbane, Australia	27.3 °S	4.0	71	3	4.5
Kathmandu, Nepal	27.4 °N	2.0	79	28	11
Gainesville, Florida, USA	29.4 °N	0.2	60	3	–
Tel Aviv, Israel	32.1 °N	3.0	67	15	3.2
Tokyo	35.7 °N				
Nilgata	37.9				
Toronto, Canada	43.4 °N	1.0	58	2	3.3
China	50.4				
Berlin, Germany	52.3 °N	5.0	50	12	0.73
Uppsala, Sweden	59.9 °N	4.4	60	9	0.66

significant difference in the return strokes after about 30 μs from the start of the stroke. There are bursts of a leader-like pulse activity in the positive return stroke. Positive return strokes have slow field variation and hence give the relation between these slow field variations and upper atmospheric lightning activities such as sprites and jets. The positive lightning ground flashes produce long impulse current, leading to the upper atmospheric phenomena and high frequency radiations (Gomes & Cooray, 1998).

The middle and upper atmosphere, known as sprites and jets, have relations with positive return strokes that carry the highest lightning currents up to 300kA and a large amount of charge transfer to the ground (Rakov & Uman, 2003). Kong et al. (2008), who used the high-speed video camera and electric field changes from slow as well as fast antenna, have reported the optical and electric field observations of a positive stroke based on both the images. The difference between positive and negative lightning flash was reported by Nag and Rakov (2012) in terms of leader propagation mode, the occurrence of continuing currents, branching, and multiplicity. Positive lightning flashes have more intense damage than negative lightning flashes (Schumann et al., 2013). The positive lightning flashes are the dominant type and are frequently found in the winter cold season and dissipating stage of thunderstorms

(Nag & Rakov, 2014). The upward connecting leader in positive ground flashes and downward positive leaders have the same electric field waveforms, as explained by Johari et al. (2016). As already mentioned in Section 2.3, the tripole structure of a thundercloud, Nag and Rakov (2012) have explained that the positive cloud to ground flashes occur rarely. Hence, they are less studied and less understood. However, the charge structure of the cloud remains a mystery.

Positive ground flashes are related to upper atmospheric discharges such as sprites, halos, jets, etc. and severe weather phenomenon such as tornadoes, hails, derecho, etc. Hence, negative ground flashes are predominant during the normal thunderstorms explained by Price (2008, 2009). They also explained the lightning thunderstorm, the dynamics of thunderstorm, and its internal process. About 60% of all cloud to ground lightning flashes, the positive cloud to ground lightning flashes occurred during an episode of 5 tornadoes within 1 h (Carey et al., 2003). In severe windstorms with straight-line winds, called a derecho, Price and Murphy (2002) found the positive CG lightning activity during the most intense part of the storm.

Adhikari and Sharma (2020) reported that the signatures of lightning electromagnetic fields wave related to the Himalayan country Nepal have a different structure. The Himalayan mountainous country consists of three regions, the Terai region, hilly region, and Himalayan region. The Himalayan region lies in the country's northern part, and the Terai region lies in the southern part. Of the ten tallest mountains in the world, eight are located in the Himalayan region of Nepal. The tallest mountain, Mount Everest, is at 8848.86 m and the lowest height in the Terai region is about 60 m with a range of 160 km from south to north part of the country. Atmospheric structure and hydrometeorological processes along the south slopes of the Himalayas are significant for the lightning phenomena. Hence, lightning is of much interest to the scientific community for the rugged terrain. The development of the thundercloud, signatures of lightning flashes, and its charge structures are also equally interesting. The parameters of microsecond-scale electric and magnetic field waveforms produced by positive ground flashes are represented in Table 3.7. They are compared with those from different geographical regions, which is represented in Table 3.7.

Adhikari et al. (2016) reported that most of the positive cloud to ground lightning flashes were single return strokes, as already mentioned; however, some of the positive cloud to ground lightning flashes consisted of two or more strokes. Among the number of positive cloud to ground flashes recorded and analyzed, the average number of strokes per flash, called multiplicity, is 1.1. The percentage of single-stroke flashes is 91.7%, and up to four strokes were observed in a single flash. An example of a four-stroke flash has been depicted in Figure 3.8 (a), and the expansion of these strokes are given in (b), (c), and (d), respectively.

Adhikari (2019a) also reported that some of the positive cloud to ground lightning flashes consisted of return strokes with multiple peaks, called doubly peaked strokes. Due to the surge of the residual charge in the phenomena of lightning discharge along the ionized channel, multiple peaks can be viewed, which is given in Fig. 3.9a. As already mentioned, the negative cloud to ground lightning flashes have stepped leaders and branching, but these are lacking in the positive cloud to ground lightning

Table 3.7 Parameters of microsecond-scale electric and magnetic field waveforms produced by positive ground flashes

Parameter			Location	Sample size	Mean (μs)	Range (μs)
Initial electric field peak normalized to 100 km, V/m	Cooray and Lundquist (1982)		Sweden	58	11.5	4.5–24.3
	Cooray et al. (1986)		Denmark	22	13.9	6.5–26
Zero to peak rise time (μs)	Rust et al. (1981)		USA	15	6.9	4–10
	Cooray and Lundquist (1982)		Sweden	64	13	5–25
				52	12	5–25
	Cooray (1986)		Sweden	20	8.9	4–12
	Ishii and Hojo (1989)	Summer	Japan	32	13.2	6–22
		Winter		123	21.2	8–44
	Ushio et al. (1998)	Winter	Japan	19	18	
10–90 percent rise time	Cooray (1986)		Sweden	15	6.2	3–9
	Hojo et al. (1985)	Summer	Japan	44	6.7	
		Winter		32	8.7	
	Adhikari et al. (2016)		Nepal	146	9.44	2.41–44.97
Zero crossing time	Ishii and Hojo (1989)	Summer	Japan	34	151	80–280
		Winter		89	93	30–160
	Adhikari et al. (2016)		Nepal	146	60.45	12.49–460.3

flash. However, some of the positive cloud to ground lightning flashes consist of up to four strokes, as already mentioned in chapter 2.5, and the example of the Figure is shown in Fig. 2.2.

The rise time (t_r) of electric field change from 10% to 90% of the maximum field of the return stroke is 9.44 μs, and the zero crossing time between the rise of the electric field from the reference level and decrease of the field to the reference level is 60.45 μs. Similarly, The FWHM, time duration between the 50% rise and 50% fall electric field pertinent to a return stroke from the reference level is 14.18 μs. The correlation between the rise time and zero crossing time is found to be 0.48, between rise time and FWHM is 0.54 and between zero crossing time and FWHM is 0.62. The positive ground flashes occurred almost over the hilly regions. An example of selected data from the whole dataset of all lightning flashes is shown in Fig. 3.9b (Adhikari et al., 2016).

The comparison with the data from other geographical regions is represented in Table. Qie et al. (2002) analyzed the 23 positive cloud to ground lightning flashes and reported that 87% were single stroke and 13% of the total flashes were multiple strokes with multiplicity 1.13. Analyzing the 204 positive cloud to ground lightning flashes by Fleenor et al. (2009), it was observed that 96% were single-stroke, and the 4% had two strokes with multiplicity an average stroke per flash of 1.04. A total of

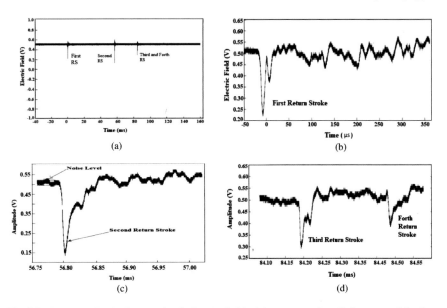

(a)

(b)

(c)

(d)

Fig. 3.8 An example of a four-stroke flash recorded in (**a**), an expansion of signature of the first return stroke of the four-stroke positive flash in (**b**), an expansion of signature of the second return stroke of the same positive flash in (**c**) and the expansion of signature of the third and fourth return stroke of the four-stroke positive flash is shown in (**d**)

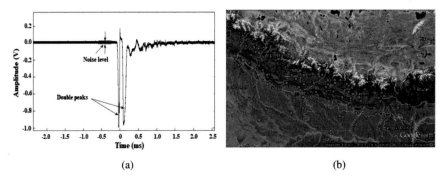

(a)   (b)

Fig. 3.9 An example of the waveform of a positive lightning return stroke with multiple peaks is shown in (**a**) and occurrence of positive lightning activities over the Himalayan regions is shown in (**b**)

133 positive cloud to ground lightning flashes were analyzed by Saba et al. (2010), and it was reported that 81% were single stroke and 19% were multiple strokes with a multiplicity of 1.2 and the mean inter-stroke interval 94 ms. Similarly, Qie et al. (2013) in Da Hinggan ling, China, analyzed 185 cloud to ground lightning flashes and reported about 95% to be single-stroke CG flashes with a multiplicity of 1.06 and inter-stroke interval of 97.83 ms in average and rise time of 7.77 μs, whereas

Table 3.8 Comparison of positive CG multiplicity, rise time, etc. from different geographical regions

Researcher	Location	No. of positive CG flashes	Ratio of single strokes	Strokes/flash	Inter-stroke interval (ms)	Rise time (μs)	Zero crossing time (μs)
Adhikari et al. (2016)	Kathmandu, Nepal	133	91.7%	1.1	33.77	9.44	60.45
Qie et al.(2013)	Da Hinggaan ling, china	185	94.6%	1.06	97.83	7.77	–
Fleenor et al. (2009)	America	204	96%	1.04	–	–	–
Qie et al.(2002)	Gansu, China	23	87%	1.13	88.2	–	–
Saba etal. (2010)	Austria, Brazil, USA	103	81%	1.2	94	–	–
Nag and Rakov (2012)	Florida	52	81%	1.2	77	–	–
Hojo et al.(1985)	Tokyo Nilgata	44 32	– –	– –	– –	6.7 8.7	– –
Schumann et al. (2013)	Brazil	(72)	–	–	–	5.7	–
Cooray (1986)	Sweden	(15) (20)	– –	– –	– –	6.2 8.9	– –

91.7% of single-stroke had a multiplicity of 1.1, and average inter-stroke interval was 33.77 ms as reported by Adhikari et al. (2016). All the characteristics of positive cloud to ground lightning flashes are shown in Table 3.8. Adhikari et al. (2016), by analyzing 133 positive cloud to ground lightning flashes, reported the rise time to be 9.44 μs with the system of sampling interval 3.2 ns. As shown in Table 3.8, Hojo et al. (1985) found the rise time of field change of return stroke in Nilgata to be 8.7 μs and at Tokyo to be 6.7 μs. Schumann et al. (2013) analyzed 72 positive cloud to ground lightning flashes and found the rise time of 5.7 μs. Cooray (1986) analyzed 15 and 20 positive cloud to ground lightning return strokes and found the average rise time to be 6.2 μs and 8.9 μs, respectively in Sweden. On comparing the results, the propagation effect over the rugged hills is an important effect of the higher rise time of the electric field change. However, at Nilgata and Uppsala, the rise time of the return strokes were 8.7 μs and 8.9 μs, respectively, which are higher. These data were taken from 0 to 100% of the peak amplitude of the return stroke and may have some propagation effect over land.

Baral and Mackerras (1993) reported the positive cloud to ground lightning flashes were frequently observed and recorded relatively high, as much as 38% during the post-monsoon. As shown in Fig. 3.10, the distance between the negative charge of

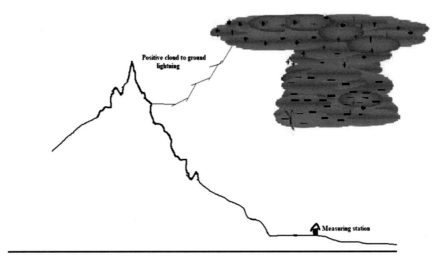

Fig. 3.10 Comparing the distance between the charge structure in the cloud and hill to that between the ground level and the clouds

the cloud and ground is higher in comparison with the distance between the positive charge of the cloud and the hills. Hence, positive cloud to ground lightning occurs frequently. Not only this, Adhikari et al. (2016) reported that the inter-stroke interval observed in Kathmandu, Nepal, is found to be relatively shorter than those observed in other locations. Relatively, the distance between the hills and the clouds is small compared to the distance between the sea level and the clouds (Adhikari, 2019b). As shown in this Fig. 3.10, the average rise time of the positive cloud to ground return strokes is found to be longer than those observed at other locations, as mentioned in Table 3.8. Also, the longer rise time to the propagation effects over the rugged hills as in the figure may be due to the unavailability of the synchronized lightning location system.

3.5 Unusual Lightning Discharge

Electrical discharge phenomena between two charged cloud regions or between the cloud and the ground are called lightning discharge phenomena. As Benjamin Franklin verified that lightning is the process of discharge phenomena, the myth of lightning was transformed into its reality. The lightning discharge phenomenon occurs between two clouds or cloud to air or cloud to ground. During this process, the charges of the cloud transfer from one to another, and hence the net amount of charge in the cloud decreases. There are many processes to form lightning, and each has different properties. The different properties of lightning discharges were described using fast optical methods, current measurement methods, measurement of stepped leaders, electromagnetic instruments, etc. Several methods were being developed, and a number of previously unrecognized lightning events also began to be identified day by day. For example, D. J. Malan in 1937 originally found the M-components in the steady continuing current, Fisher et al. (1993) gave the first report on correlating current and optical records of M-components, Rakov (1998) gave the first report on correlated current, electric field and magnetic field records of M-components in triggered lightning. Shao (1993) first studied the VHF images of M-components, N. Kitagawa and M. Kobayashi, reported first the K-changes in the lightning process, K. Bergers and coworkers (1975) first reported a well-documented study of positive lightning discharge phenomena at Monte-San Salvatore in Switzerland, Lu et al. (2013), Nag and Rakov (2010), Sharma et al. (Sharma, Fernando, et al., 2008), Smith et al. (1999) studied about the compact intra-cloud discharges (CIDs) and were observed at lower latitudes, producing the characteristic narrow bipolar pulses (NBPs). Isolated breakdown pulses (IBPs) were observed by Sharma et al. (Sharma, Cooray, et al., 2008), and similar events were reported by Nag and Rakov (2008b), and Rakov and Uman (2003) have observed transient luminous events including sprites, halos, jets, and elves in the middle and upper atmosphere.

Among the different lightning discharges, cloud to cloud or cloud to ground lightning discharge on each lightning flash is unique. For the cloud to ground lightning, Rakov and Uman (2003), Uman (1987), etc. have described that about 90 % of the total ground flashes are negative ground flashes, and less than 10 % of the total ground flashes are positive ground flashes. Similarly, some common features of cloud flashes are also established by various researchers. Kitagawa and Brook (1960) classified cloud flash into three stages, as mentioned in chapter 3.3. However, Bils et al. (1988), Villanueva et al. (1994) classified the cloud to cloud lightning flashes into two stages: the active stage as the first stage of the cloud and the second or final stage. Sharma et al. (2005), Adhikari and Sharma (2020), etc. also categorized the cloud to cloud lightning into two stages, active stage exhibiting large microsecond-scale bipolar pulses and final stage being accompanied by smaller pulses as already discussed in the previous chapter. There are more types of lightning flashes, such as bipolar lightning flashes, unusual lightning flashes, etc. and there are different processes generally in a negative lightning flash, including preliminary breakdown process (PB), stepped leader process, first return stroke, dart leader, subsequent return stroke, continuing current, M-component, J process, K-changes, and so on.

Adhikari, et al. (2017) explained that the opposite polarity pulse prior to the main waveform here is called unusual events in lightning. Generally, the stepped leaders are moving along the same direction as the main events, as shown in Fig. 2.2. However, in these waveforms, the leader pulses are in the opposite direction to the main events, as shown in Figure 3.11. The electric field signature on all these (a), (b), (c) of Figure 3.11 has opposite polarity pulses and four-stroke positive lightning flash, also shown in part (c) of this Figure 3.11, all four strokes exhibited large opposite polarity leader pulses as unusual lightning, prior to the return stroke waveform. This is another mystery related to lightning, as we know that positive flashes are generally composed of a single stroke. They also explained that these unusual lightning events were observed on a single day in Sweden but frequently in Kathmandu, Nepal.

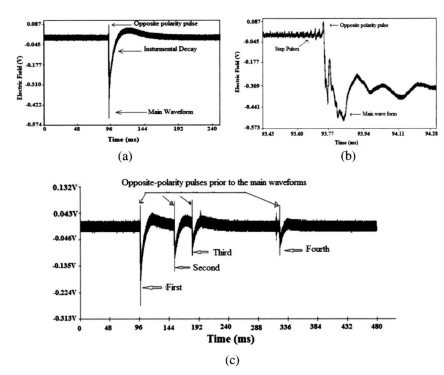

Fig. 3.11 An example of unusual lightning event in which **a** the electric field waveform shown on a 240 ms time scale. **b** The expansion of unusual lightning event (a) expanded on a 850 s time scale. **c** Electric field signature of what appears to be an unusual four-stroke positive lightning flash in which all four strokes exhibited large opposite polarity pulses prior to the main waveform

Adhikari et al. (2017) reported the different characteristics of electric field signatures produced by the unusual lightning events recorded in Kathmandu during the premonsoon period within 500 km. The atmospheric electricity sign convention is used in this measurement of lightning electric fields pertinent to the unusual lightning events. The measurement campaign was conducted in Kathmandu, about 1300 m above sea level, and the lightning wave signatures were observed, recorded, and analyzed. In this campaign, mainly positive cloud to ground lightning, unusual waveforms, and cloud to cloud lightning waveforms were observed and recorded. The vertical electric field measured with a parallel plate antenna system was mounted on the top of a 12 m height building to measure the lightning waveform. They also explained the instruments to measure the vertical electric fields, 60pF capacitance of parallel plate antenna connected with buffer amplifier and Picoscope 6404D through RG-58 coaxial cable. The RC decay time of 13 ms is sufficient to record the radiation components of the lightning electric field. Only one case of positive cloud to ground flash having the opposite polarity leader pulses just before the main waveform was observed, which is called an unusual lightning waveform reported by Johari et al. (2016) in Sweden. These opposite polarity leader pulses are reported by them, then the bidirectional leader inside the cloud due to negatively charged leader branch propagates toward the measurement station.

Uman (2001) explained the coaxial cable, antenna systems, and electronic circuits associated with them. The horizontal coaxial cables act as a capacitor in the circuit. There are other capacitor components such as capacitance between the upper plate of the antenna and the ground, the capacitors used in the circuits, and the coaxial cable between them. The use of a coaxial cable may be for coupling with other components and maybe to lead the unwanted horizontal electric field to distort the main waveform. According to Adhikari et al. (2017), when a downward-directed positive leader approaches the ground, the electric field near the grounded objects increase. Due to this effect, the negative leaders will launch in the upward direction and will try to intercept the descending positive leaders. There may be several such types of upward-directed negative electric field leaders, and one of those upward leaders will succeed to contact. Other upward-directed negative electric field leaders will collapse at the time of return stroke. These types of upward-directed negative electric field leaders are called unsuccessful upward negative leaders. The current related to the motion of negative charge of unsuccessful upward negative leaders back to the ground produces a positive electric field change. The associated current, which produces the electric field change opposite to that of the positive return stroke, results in the collapse of the unconnected upward negative leaders.

An example of this type of electric field waveform is shown in Fig. 3.11. Hence in this scenario, there may be multiple times within the same flash, as shown in Fig. 3.11c. In this figure, the opposite polarity step pulses are seen prior to the large main event of opposite polarity pulses. This type of unusual lightning event was first observed and recorded by the electric field measurements systems in Sweden. Similarly, the same vertical electric field measurement with a parallel plate antenna system was used in Kathmandu, Nepal, and the unusual lightning electric field waveform was reported by Adhikari et al. (2017). The example of this electric field waveform signature of an unusual lightning event is shown in Fig. 3.12. In this Figure, electric field signatures of a single stroke are given in (a) and multiple strokes of the unusual lightning events (b).

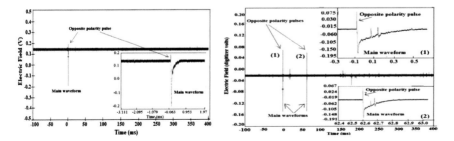

Fig. 3.12 Examples of electric field signatures of unusual lightning events **a** the single stroke of unusual lightning events and **b** electric field signatures of multiple strokes (Adhikari et al., 2017)

The parameters of the rise time of the opposite polarity leader pulses (**T1**), total duration of opposite polarity leader pulses (**T2**), the rise time of the main waveform (**T3**), total duration of the main waveform (**T4**), the amplitude of the opposite leader pulse (**A1**), and amplitude of the main waveforms (**A2**) were used by Adhikari et al. (2017) and analyzed the electric field waveforms recorded in Uppsala and Kathmandu and then compared. The parameters of the waveform are represented in Fig. 3.13. The parameters of field waveforms produced by unusual lightning in both locations were reported by Adhikari et al. (2017), which is adopted here in Table 3.9.

Adhikari et al. (2017) explained that lightning activities of Uppsala and Kathmandu were compared because of the characteristics of similar unusual activities,

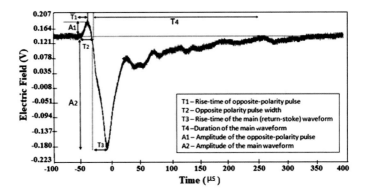

Fig. 3.13 Definitions of the field waveform parameters adopted from Adhikari et al. (2017)

Table 3.9 Comparison of the parameters of unusual electric field waveforms produced by lightning events observed in two locations (Adhikari et al., 2017)

Parameters	Location	Maximum	Minimum	Mean	S. D	C.V. = (S.D./MEAN)
T1	A	9.00 μs	1.25 μs	4.21 μs	1.96 μs	46.56
	B	20.71 μs	0.40 μs	3.87 μs	2.77 μs	71.58
T2	A	13.09 μs	3.62 μs	7.97 μs	2.26 μs	28.36
	B	36.01 μs	1.0 μs	10.77 μs	6.09 μs	56.55
T3	A	101.86 μs	4.79 μs	57.09 μs	32.04 μs	56.12
	B	147.5 μs	1.94 μs	39.71 μs	33.64 μs	84.71
T4	A	11.15 ms	4.17 ms	8.24 ms	1.55 ms	18.81
	B	2.292 ms	90.22 μs	521 μs	329.93 μs	63.31
T4/T2	A	2229.28	582.09	1109.71	366.85	33.06
	B	363.2	6.15	68.04	62.07	91.23
T3/T1	A	63.22	1.42	15.77	11.4	72.29
	B	93.78	0.47	12.15	11.18	92.02
A2/A1	A	7.25	1.11	3.05	1.3	42.62
	B	13.93	0.64	2.65	1.62	61.13

A stands for Uppsala and B stands for Kathmandu

as shown in Table 3.9. They compared the features of waveforms of both locations. Adhikari et al. (2017) explained that by comparing the characteristics of the
analyzed waveforms, no significant difference was found. From Table 3.9, the mean
of the main waveform duration in Sweden was 8.24 ms, which was significantly
longer than those of 521 μs in Kathmandu. They also concluded that the average
distance between the lightning events from the measuring station in Sweden was
nearer than in Nepal. Hence, the static and induction electric field components were
strongly attenuated in Nepal. To attenuate the static and inductive field, the rugged
terrain played a significant role.

3.6   Other Discharges

The means by which the amount of accumulated charge decreases is called discharge.
As mentioned in previous chapters, there are various methods of discharge that can
be seen by analyzing the waveform of the lightning discharge. Besides the cloud
to cloud lightning, positive cloud to ground lightning, negative cloud to ground
lightning, unusual lightning, there are many other types of lightning such as rocket
lightning, ribbon lightning, bead lightning, ball lightning, warm cloud lightning, etc.
(Rakov & Uman, 2003; Uman, 2001).

3.6.1   Warm Cloud Lightning

This lightning is produced by the thundercloud when its temperature becomes well
below freezing level, i.e., the temperature below 0 °C. As already mentioned in
Chapter 1, the temperature of the cloud decreases as the height of the cloud increases.
If the temperature of the cloud is below the freezing level, lightning occurs due to
the motion of super-cooled water droplets and ice particles inside the thunderstorm.
If the height of the cloud is at the lower altitude which is below the freezing level,
the cloud is at the temperature above 0 °C. Lightning that occurs when the clouds
are at a lower altitude, lower than the freezing level, is called warm cloud lightning.

3.6.2   Rocket Lightning

It is the cloud to air discharges in the form of a rocket, propagation in a slow horizontal
manner within a time interval in the order of 10 ms. The discharges are from cloud to
air, outside the cloud, so, it can be seen clearly. It may likely be due to the intra-cloud
discharges, but the discharge channel is visible outside the cloud; one example of
rocket lightning is shown in Fig. 3.13a.

3.6.3 Ribbon Lightning

When the cloud to ground lightning occurs, the discharge channel is shifted by the wind, then the strokes in the flashes are separated horizontally, which increases the width and forms the shape of ribbon called ribbon lightning.

3.6.4 Bead Lightning

For the cloud to ground lightning discharges, it is not a continuous channel, i.e., it breaks up or appears to break up in between, some tens of meters in length. Then, the channel can be seen as a bead. This type of bead formation in the cloud to ground lightning is called bead lightning. Bead lightning has a long continuing current in the rocket-initiated triggered lightning. One example of bead lightning is shown in Fig. 3.13b.

3.6.5 Ball Lightning

It is also formed in the cloud to ground lightning discharges. Ball lightning is luminous, red, and often appears in the shape of a ball, from an orange to a large basketball. The duration of this type of lightning is about 1 s reported by Uman (2001). The ball lightning is generally more horizontally near the ground, maintained at a roughly constant brightness. Examples of ball lightning are shown in Fig. 3.13c and d (Fig. 3.14).

(a)   (b)

(c)   (d)

Fig. 3.14 Different lightning **a** an example of rocket lightning, **b** an example of bead lightning **c** and **d** are examples of ball lightning

References

Adhikari, P. B. (2019a). Investigations of the features of electromagnetic fields due to lightning measured in Kathmandu. Ph.D. Thesis, Tribhuvan University.

Adhikari, P. B. (2019b), Various types of lightning electric field signatures observed in Kathmandu, Nepal. *Journal of Astrophysics & Aerospace Technology 7*(2), 164. https://doi.org/10.4172/2329-6542

Adhikari, P. B., & Sharma, S. R. (2020). Characteristic features of electric fields radiated by cloud flashes in Himalayan region. *International Journal of Antennas and Propagation, 10*(1155), 1–8.

Adhikari, P. B., & Adhikari, A. (2021), Comparing the wave characteristics of breakdown pulses of the lightning waveforms in the Himalayan Region. *The Scientific World Journal, 2021*. https://doi.org/10.1155/2021/6381439

Adhikari, P. B., Sharma, S. R., & Baral, K. N. (2016). Features of positive ground flashes observed in Kathmandu, Nepal. *Journal of Atmospheric and Solar-Terrestrial Physics, 145*, 106–113.

Adhikari, P. B., Sharma, S. R., Baral, K. N., & Rakov, V. A. (2017). Unusual lightning electric field waveforms observed in Kathmandu, Nepal, and Uppsala, Sweden. *Journal of Atmospheric and Solar-Terrestrial Physics, 164*, 172–184.

Andrews, C. J., Cooper, M. A., Darveniza, M., and Mackerras, D. (1992). *Lightning injuries: Electrical, medical and legal aspects.* CRC.

Baral, K., & Mackerras, D. (1993). Positive cloud-to-ground lightning discharges in Kathmandu thunderstorm. *Journal of Geophysical Research, 98*, 10331–10340.

Bazelyan, E., & Raizer, Y. (2000). *Lightning physics and lightning protection.* Institute of Physics Publishing Dirac House.

Berger, K. (1977). *The earth flash, in lightning, physics of lightning.* Academic Press.

Berger, K., Anderson, R. B., & Kroninger, H. (1975). Parameters of lightning flashes. *Electra, 80*, 223–237.

Berger, K., & Vogelsanger, E. (1966). Photographische blitzuntersuchungen der jahre 1955-1963 auf dem monte sen salvatore. *Bull. Schweiz. Elektrotech. Ver., 57*, 599–620.

Bils, J., Thomson, E., Uman, M., & Mackerras, D. (1988). Electric field pulses in close lightning cloud flashes. *Journal of Geophysical Research, 93*, 15933–15940.

Bitzer, P. M., Christian, H. J., Stewart, M., Burchcfield, J., Podgorny, S., Corredor, D., Hall, J., Kuznetsov, E., & Franklin, V. (2013). Characterization and applications of VLF/LF source locations from lightning using the Huntsville Alabama Marx Meter Array. *Journal of Geophysics Research: Atmospheres, 118*, 3120–3138.

Brook, M. (1992). Breakdown electric fields in winter storms. *Research Letters in Atmospheric Electricity, 12*, 47–52.

Carey, L., Peterson, W., and Rutledge, S. (2003). Evolution of cloud to ground lightning and storm structure in spencer, South Dakota, Tornadic super cell of 30 May 1998 (Vol. 131).

Clarence, N., & Malan, D. (1957). Preliminary discharge processes in lightning flashes to ground. *Quarterly Journal of the Royal Meteorological Society, 83*, 161–172.

Cooray, V. (1984). Further characteristics of positive radiation fields from lightning in Sweden. *Journal of Geophysical Research, 84*, 11807–11815.

Cooray, V. (1986). A novel method to identify the radiation fields produced by positive return strokes and their sub-microsecond structure. *Journal of Geophysical Research, 91*, 7907–7911.

Cooray, V., & Lundquist, S. (1982). On the characteristics of some radiation fields from lightning and their possible origin in positive ground flashes. *Journal of Geophysical Research, 87*, 11203–11214.

Cooray, V., & Lundquist, S. (1985). Characteristics of the radiation fields from lightning in Sri Lanka in the tropics. *Journal of Geophysical Research, 90*, 6099–6109.

Fisher, R. J., Schnetzer, G. H., Thottappillil, R., Rakov, V. A., Uman, M. A., & Goldberg, J. D. (1993). Parameters of triggered-lightning flashes in Florida and Alabama. *Journal of Geophysical Research, 98*, 22887–22902.

Fleenor, S., Biagi, C., Cummins, K., Krider, E., & Shao, X. (2009). Characteristics of cloud to ground lightning in warm-season thunderstorms in the Great Plains. *Journal of Geophysical Research, 91*, 333–352.

Fuquay, D. M. (1982). Positive cloud-to-ground lightning in summer thunderstorms. *Journal of Geophysical Research, 87*, 7131–7140.

Golde, R. H. (1947). Occurrences of upward streamers in lightning discharges. *Nature, 160*, 395–396.

Gomes, C., & Cooray, V. (1998). Long impulse currents associated with the positive return strokes. *Journal of Atmospheric and Solar-Terrestrial Physics, 60*, 693–699.

Gomes, C., Cooray, V., Fernando, M., Montano, R., & Sonnadara, U. (2004). Characteristics of chaotic pulse trains generated by lightning flashes. *Journal of Atmospheric and Solar-Terrestrial Physics, 66*, 1733–1743.

Gomes, C., and Cooray, V. (2001). Characteristics of cloud flashes. In *14th International Zurich symposium, EMC 58J3*.

Hayenga, C. O. (1984). Characteristics of lightning VHF radiation near the time of return strokes. *Journal of Geophysical Research, 89*, 1403–1410.

Hojo, J., Ishii, M., Kawamura, T., Suzuki, F., & Funayama, R. (1985). The fine structure in the field change produced by positive ground strokes. *Journal of Geophysical Research, 90*, 6139–6143.

Ishii, M., & Hojo, J.-I. (1989). Statistics on fine structure of cloud-to-ground lightning field waveforms. *Journal of Geophysical Research, 94*, 13267–13274.

Isikawa, H. (1961). Nature of lightning discharges as origins of atmospherics. *Proceedings of Research Institute Atmosphere (nagoya University), 8A*, 1–273.

Johari, D., Cooray, V., Rahman, M., Hettiarachchi, P., & Ismail, M. (2016). Characteristics of leader pulses in positive ground flashes in Sweden. *Electric Power System Research, 153*, 3–9.

Karunarathne, S., Marshall, T. C., Stolzenburg, M., Karunarathna, N., Vickers, L. E., Warner, T. A., & Orville, R. E. (2013). Locating initial breakdown pulses using electric field change network. *Journal of Geophysics Research: Atmospheres, 118*, 7129–7141.

Kitagawa, N., and Kobayashi, M. (1959). Field changes and variations of luminosity due to lightning flashes. In Recent advances in atmospheric electricity (L. G. Smith ed., pp. 485–501). Pergamon.

Kitagawa, N., & Brook, M. (1960). A comparison of intra-cloud and cloud-to-ground lightning discharges. *Journal of Geophysical Research, 65*, 1189–1201.

Kitagawa, N. (1957). On the electric field change due to the leader process some of their discharge mechanism. *Papers in Metereological Geophysics (Tokyo), 7*, 400–414.

Kobayashi, M., Kitagawa, N., Ikeda, T., & Sato, Y. (1958). Preliminary studies of variation of luminosity and field change due to lightning flashes. *Papers in Meteorological Geophysics (Tokyo), 8*, 29–34.

Kong, X., Qie, X., & Zhao, Y. (2008). Characteristics of downward leader in a positive cloud-to-ground lightning flash observed by high-speed video camera and electric field changes. *Geophysical Research Letters, 35*(L05816), 1–5.

Krehbiel, P. R. (1981). *An analysis of electric field changeproduced by lightning.* University of Manchester Institute of Science and Technology.

Krehbiel, P. R., Brook, M., & McCrory, R. A. (1979). An analysis of the charge structure of lightning discharges to ground. *Journal of Geophysical Research, 84*(C5), 2432–2456.

Krider, E. P., Radda, G. J., & Noggle, R. C. (1975). Regular radiation field pulses produced by intra-cloud lightning discharges. *Journal of Geophysical Research, 80*, 3801–3804.

Le Vine, D. M. (1980). Sources of the strongest RF radiation from lightning. *Journal of Geophysical Research, 85*, 4091–4095.

Liu, X. S., & Krehbiel, P. R. (1985). The initial streamer of intra-cloud lightning flashes. *Journal of Geophysical Research, 90*, 6211–6218.

Lu, F., Zhu, B., Zhou, H., Rakov, V. A., Xu, W., and Qin, Z. (2013). Ob-servation of compact intra-cloud lightning discharges in the northern region (51°) N of China: Journal of Geophysical Research: Atmospheres; 118; 4458 - 65.

Mac Gorman, D., & Rust, W. (1998). *The electrical nature of storms.* Oxford University Press.

Mackerras, D. (1968). A comparison of discharge processes in cloud and ground lightning flashes. *Journal of Geophysical Research, 73*, 1175–1183.

Mackerras, D., and Darveniza, M. (1994). Latitudinal variation of lightning occurrence characteristics. *Journal of Geophysical Research, 99*, 6195–6201.

Mazur, V. (1989). Triggered lightning strikes to aircraft and natural intracloud discharges. *Journal of Geophysical Research, 94*, 3311–3325.

Mazur, V., Gerlach, J. C., & Rust, W. D. (1984). Lightning flash density versus altitude and storm structure from observations with uhf- and s-band radars. *Geophysical Research Letter, 11*, 61–64.

Nag, A., & Rakov, V. A. (2008a). Pulse trains that are characteristics of preliminary breakdown in cloud-to-ground lightning but are not followed by return stroke pulses. *Journal of Geophysical Research, 113*(D01102), 1–12.

Nag, A., and Rakov, V. A. (2008b). An experimental study of electric field pulses produced by cloud and ground lightning discharges. In *19th International Zurich symposium on electromagnetic compatibility*, Singapore.

Nag, A., & Rakov, V. A. (2009). Some inferences on the role of lower positive charge region in facilitating different types of lightning. *Geophysical Research Letters, 36*(L05815), 1–05.

Nag, A., & Rakov, V. A. (2010). Compact intracloud lightning discharges: Mechanism of electromagnetic radiation and modeling. *Journal of Geophysical Research, 115*(D20102), 1–20.

Nag, A., & Rakov, V. A. (2012). Positive lightning: An overview, new observations, and inferences. *Journal of Geophysical Research: Atmospheres, 117*(D08109), 1–20.

Nag, A., & Rakov, V. A. (2014). Parameters of electric fields waveforms produced by positive lightning return strokes. *IEEE—TEMC, 56*, 932–939.

Nag, A., DeCarlo, B. A., & Rakov, V. A. (2009). Analysis of microsecond- and submicrosecond-scale electric field pulses produced by cloud and ground lightning discharges. *Atmospheric Research, 91*, 316–325.

Ogawa, T., & Brook, M. (1964). The mechanism of the intracloud lightning discharge. *Journal of Geophysical Research, 69*, 514–519.

Ogawa, T. (1995). *Lightning currents, handbook of atmospheric electrodynamics*. CRC.

Pierce, E. T. (1955). The development of lightning discharges. *Quartly Journal of Royal Meterological Society, 81*, 229–240.

Price, C. (2008). Lightning sensors for observing, tracking and now casting severe weather. *Sensors, 8*, 157–170.

Price, C., Asfur, M., & Yair, Y. (2009). Maximum hurricane intensity preceded by increase in lightning frequency. *Nature Geoscience, 2*, 329–332.

Price, C., & Murphy, B. (2002). Lightning activity during the 1999 superior derecho. *Geophysical Research Letter, 29*, 57.1–57.4.

Proctor, D. E. (1971). A hyperbolic system for obtaining vhf radio pictures of lightning. *Journal of Geophysical Research, 76*, 1478–1489.

Proctor, D. E. (1981). VHF radio pictures of cloud flashes. *Journal of Geophysical Research, 86*, 4041–4071.

Proctor, D. E. (1997). Lightning flashes with high origins. *Journal of Geophysical Research, 102*, 1693–1706.

Qie, X., Yu, Y., Wang, D., Wang, H., & Chu, R. (2002). Characteristics of cloud to ground lightning in Chinese inland plateau. *Journal of Meteorological Society, 80*, 745–754.

Qie, X., Wang, Z., Wang, D., & Liu, M. (2013). Characteristics of positive cloud to ground lightning in da Hinggan Ling forest at relatively high latitude, North-Eastern China. *Journal of Geophysical Research: Atmospheres, 118*, 13393–13404.

Rakov, V. A. (1998). Some inferences on the propagation mechanisms of dart leaders and return strokes. *Journal of Geophysical Research, 103*, 1879–1887.

Rakov, V. A., & Uman, M. A. (2003). *Lightning: Physics and effects*. Cambridge University Press.

Rhodes, C. (1989). Interferometric observations of VHF radiation of lightning. New Mexico Institute of Minerals and Technology, Socorro.

Richard, P., & Auffray, G. (1985). VHF-UHF interferometric measurements, applications to lightning discharge mapping. *Radio Science, 20*, 171–192.

Rison, W., Thomas, R. J., Krehbiel, P. R., Hamlin, T., & Harlin, J. (1999). A GPS-based three-dimensional lightning mapping system: Initial observation in central New Mexico. *Geophysical Research Letters, 26*, 3573–3576.

Rust, W. D., MacGorman, D. R., & Arnold, R. T. (1981). Positive cloud to ground lightning flashes in severe storms. *Geophysical Research Letter, 8*, 791–794.

Saba, M. M. F., Schulz, W., Warner, T. A., Campos, L. Z., Schumann, C., & Krider, E. P. (2010). High speed video observations of positive lightning flashes to ground. *Journal of Geophysical Research, 115*(D24201), 1–9.

Schumann, C., Saba, M. M. F., DaSilva, R. B., & Schulz, W. (2013). Electric fields changes produced by positive cloud-to-ground lightning flashes. *Journal of Atmospheric and Solar-Terrestrial Physics, 92*, 37–42.

Shao, X. M., & Krehbiel, P. R. (1996). The spatial and temporal development of intracloud lightning. *Journal of Geophysical Research, 101*(26), 641–668.

Shao, X. M. (1993). *The development and structure of lightning discharges observed by vhf radio interferometer*. New Mexico Institute of Minerals and Technology, Socorro.

Sharma, S. (2007). *Electromagnetic fields radiated by lightning in tropical and temperate regions*. Ph.D. Thesis, Faculty of Science, University of Colombo, Sri Lanka.

Sharma, S. R., Fernando, M., & Gomes, C. (2005). Signatures of electric field pulses generated by cloud flashes. *Journal of Atmospheric and Solar-Terrestrial Physics, 67*, 413–422.

Sharma, S. R., Fernando, M., & Cooray, V. (2008). Narrow positive bipolar radiations from lightning observed in Sri Lanka. *Journal of Atmospheric and Solar-Terrestrial Physics, 70*, 1251–1260.

Sharma, S. R., Cooray, V., & Fernando, M. (2008). Isolated breakdown activity in swedish lightning. *Journal of Atmospheric and Solar-Terrestrial Physics, 70*, 1213–1221.

Smith, D., Shao, X., Holden, D., Rhodes, C., Brook, M., & Krehbiel, P. (1999). A distinct class of isolated intra cloud lightning discharges and their associated radio emissions. *Journal of Geophysical Research, 104*, 4189–4212.

Takagi, M. (1961). The mechanism of discharges in a thundercloud. *Proceedings of Research Institute Atmosphere (nagoya University), 88B*, 1–105.

Takagi, M., & Takeuti, T. (1963). Atmospheric radiation from lightning discharge. Proceedings of Research Institute Atmosphere, Nagoya University, 10, 1–11.

Thomas, R. J., Krehbiel, P. R., Rison, W., Hamlin, T., Boccippio, D. J., Goodman, S. J., & Christia, H. J. (2000). Comparison of ground based 3-dimensional lightning mapping observations with satellite-based LIS observations in Oklahoma. *Geophysical Research Letters, 27*, 1703–1706.

Thomson, E. M. (1985). A theoretical study of electrostatic field wave shaves from lightning leaders. *Journal of Geophysical Research, 27*, 8125–8135.

Uman, M. A. (1987). The lightning discharge (p. 377). Academic Press.

Uman, M. A. (2001). *The lightning discharge* (p. 377). Dover Publications.

Ushio, T.-O., Kawasaki, Z.-I., Matsu-Ura, K., & Wang, D. (1998). Electric fields of initial breakdown in positive ground flash. *Journal of Geophysical Research, 103*, 14135–14139.

Villanueva, Y., Rakov, V., & Uman, M. (1994). Microsecond-scale electric field pulses in cloud lightning discharges. *Journal of Geophysical Research, 99*, 14353–14360.

Weber, M. E., Christian, H. J., Few, A. A., & Stewart, M. F. (1982). A thundercloud electric field sounding: Charge distribution and lightning. *Journal of Geophysical Research, 87*, 7158–7169.

Weidman, C. D., & Krider, E. P. (1979). The radiation field wave forms produced by intra-cloud lightning discharge processes. *Journal of Geophysical Research, 84*, 3159–3164.

Weidman, C. D., Krider, E. P., & Uman, M. A. (1981). Lightning amplitude spectra of lightning radiation fields in the interval from 100 kHz to 20 mhz. *Geophysical Research Letter, 8*, 931–934.

Willett, J. C., Bailey, J. C., & Krider, E. P. (1989). A class of unusual lightning electric field waveforms with very strong HF radiation. *Journal of Geophysical Research, 94*, 16255–16267.

Willett, J. C., Bailey, J. C., Leteinturier, C., & Krider, E. P. (1990). Lightning electromagnetic radiation field spectra in the interval from 0.2 to 20 MHz. *Journal of Geophysical Research, 95*, 20367–20387.

Williams, E. R. (1989). The tripole structure of thunderstorms. *Journal of Geophysical Research, 94*(13), 151–167.

Williams, D. P., & Brook, M. (1963). Magnetic measurements of thunderstorm currents continuing currents in lightning. *Journal of Geophysical Research, 68*, 3243–3247.

Wu, T., Yoshida, S., Akiyama, Y., Stock, M., Ushio, T., & Kawasaki, Z. (2015). Preliminary breakdown of intracloud lightning: Initiation altitude, propagation speed, pulse train characteristics and step length estimation. *Journal of Geophysical Research: Atmospheres, 120*, 9071–9086.

Wu, T., Wang, D., & Takagi, N. (2019). Intracloud lightning flashes initiated at high altitudes and dominated by downward positive leaders. Journal of Geophysical Research: Atmospheres.

Yokoyama, S., Miyake, K., Suzuki, T., & Kanao, S. (1990). Winter lightning on Japan Sea coast—development of measuring system on progressing feature of lightning discharge. *IEEE Transactions on Power Delivery, 5*, 1418–1425.

Chapter 4
Theory of Lightning

4.1 Introduction

The lightning phenomenon is the discharging process in which the charge of the cloud transfers to the ground. It is an extremely complex electrical discharging phenomenon that produces electric and magnetic fields around it. In the atmosphere, the complex electrical discharge phenomena contain many micro- and macro-discharging process. In this process, the radiation of lightning is produced in a wide range of frequencies from very low to very high frequency, up to few hundred gigahertz. To understand the nature of lightning waveform and physics of the lightning, the basic phenomena are the electric and magnetic fields. To understand the electric and magnetic field of the lightning, mathematical expressions were derived here by different approaches. The mathematical derivation which describes the phenomena of lightning is given in this chapter.

4.2 Electrostatic Field

Let us consider amount of charge $+Q$ represents in a thundercloud at a vertical height H above from the ground. Malan (1963) considered the earth surface is a conducting plane and the curvature of the earth surface is very small on comparing the distances of the cloud from the measuring point. The electrical image of $+Q$ charge is $-Q$, as shown in the Fig. 4.1. Then, the produced electric field due to $+Q$ charge at the point P is E_1 and the electric field due to $-Q$ charge at the point P is E_2, and D be the horizontal distance from the point O to P as shown in Fig. 4.1.

The total electric field E is the vector sum of these two E_1 and E_2. i.e.,

$$E = E_1 + E_2 \tag{4.1}$$

© The Author(s), under exclusive license to Springer Nature Singapore Pte Ltd. 2022
P. B. Adhikari and A. Adhikari, *Lightning Discharges*,
SpringerBriefs in Applied Sciences and Technology,
https://doi.org/10.1007/978-981-19-1926-8_4

Fig. 4.1 Electric field E produced by charge $+Q$ at point P where OP is the surface of the ground

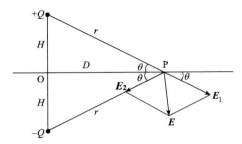

Let the angle θ as shown in figure made by vector r with the surface of the ground as shown in Fig. 4.1. Then, the total electric field is

$$E = \frac{2Q \sin \theta}{r^2} = \frac{2QH}{\left(H^2 + D^2\right)^{3/2}} \tag{4.2}$$

Substituting $H/D = x$ then $E = \dfrac{1}{D^2} \dfrac{2Qx}{\left(1 + x^2\right)^{3/2}}$. $\tag{4.3}$

On differentiating the above equation, the first derivative should be zero for the maximum electric field E, and hence we get, $1 - 2x^2 = 0$. So, $x = \frac{H}{D} = \frac{1}{\sqrt{2}}$.

If the height H of the cloud is varied, the electric field at the point P at a distance D will pass through a maximum that means distance of the lightning should be 1.4 times the height of the cloud then the electric field should be maximum. The primary charge structure in a thundercloud is dipole, in which Q_N be the amount of charge in the center of the cloud negatively and Q_P be the equal amount of positive charge in the upper part of the cloud. However, for the tripole charge structure in the cloud, there is the positive charge at the bottom of the cloud, called pocket charge (Williams, 1989). Hence in the tripole charge structure, negative charge center at the center of the cloud, equal amount of positive charge remains on the upper part of the cloud and some pocket positive charge remains on the bottom part of the cloud. Let us consider $+ Q_P$ be the amount of main positive charge, at a height H_P, $- Q_N$ be the main negative charge in equal amount centered at a height H_N. Also, D_P and DN are the horizontal distance of main positive and negative charge center respectively from the measurement station. But the pocket charge $+ q_p$ as positive is just below the main negative charge at a height of h_p, which is given in Fig. 1.1.

As Eq. 4.2, the electric field due to positive charge at the measuring point on the ground which is directed perpendicularly toward the earth is given by

$$E_+ = \frac{1}{4\pi \,\epsilon_0} \cdot \frac{2Q_P H_P}{r^3} = \frac{1}{4\pi \,\epsilon_0} \cdot \frac{2Q_P H_P}{\left(D_P^2 + H_P^2\right)^{3/2}} \tag{4.4}$$

Similarly, the electric field due to negative charge at the same point of measurement, which is directed perpendicularly to upward from the earth is given by

$$E_- = \frac{1}{4\pi \ \epsilon_0} \cdot \frac{2Q_N H_N}{(D_N^2 + H_N^2)^{3/2}} \tag{4.5}$$

Hence, the total electric field at this point is

$$E = E_+ + E_- = \frac{1}{4\pi \ \epsilon_0} \left[\frac{2Q_P H_P}{(D_P^2 + H_P^2)^{3/2}} - \frac{2Q_N H_N}{(D_N^2 + H_N^2)^{3/2}} \right] \tag{4.6}$$

As already mentioned that the equal amount of positive and negative charges i.e. $Q_P = Q_N = Q$, and are lies in vertical line. This means the horizontal distances of positive and negative charge center are at equidistant from the measuring station, i.e., $D_P = D_N = D$.

So, from the Eq. (4.6), we get,

$$E = \frac{1}{4\pi \ \epsilon_0} \left[\frac{2Q H_P}{(D^2 + H_P^2)^{3/2}} - \frac{2Q H_N}{(D^2 + H_N^2)^{3/2}} \right] \tag{4.7}$$

From this equation, it is very interesting that the features of the field is reversed in sign if the distances is increases; that means, it is negative electric field at close distance and positive electric field at far away (Adhikari & Bhandari, 2017). The distance is called reversal distance at which the sign of the field changes, suppose, D_0 be the reversal distance. Then,

$$\frac{H_P}{(D_0^2 + H_P^2)^{3/2}} = \frac{H_N}{(D_0^2 + H_N^2)^{3/2}} \tag{4.8}$$

$$D_0^2 = (H_N H_P)^{2/3} \left(H_P^{2/3} + H_N^{2/3} \right)$$

Hence, the reversal distance $D_0 = (H_N H_P)^{1/3} \left[H_P^{2/3} + H_N^{2/3} \right]^{1/2}$.

4.3 Magneto-Static Field

Let us consider a current in the return stroke channel represented in a line as shown in the Fig. 4.2. Let dz be the small elemental length along the line, which is starting from ground to the point at a height x to the point at a height of y from the ground as shown in Fig. 4.2. Let r be the distance of the elemental length dz as shown in the figure from the measuring point P that is at a horizontal distance D from point O. Within the cloud or between a cloud and ground, there is the motion of charges which constitute an electric current which is associated with a magnetic field that may be measured from the ground. Malan (1963), Uman (2001), Adhikari and Bhandari

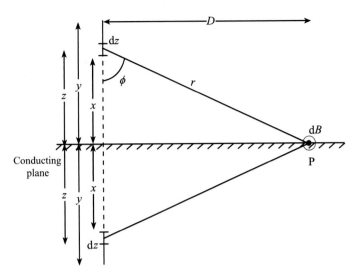

Fig. 4.2 Simple model to determine the magneto-static field at the conducting plane due to current in the return stroke channel (Uman, 2001)

(2018b) described that the magneto-static effects due to current produced by lightning discharge considering the simple model.

Let dz be the small length element in the line segment, I be the current in the line, ϕ be the angle made by r with dz. Then the magnetic field intensity (dB) at the point P on the conducting plane at a distance D is given by Biot–Savart law, which is

$$dB = \frac{\mu_0 I D dz}{4\pi (D^2 + z^2)^{3/2}} \tag{4.9}$$

To find the magnitude of the magnetic field intensity (dB) at this point at a distance D due to the current as shown in the Fig. 4.2, in a vertical line of length $y - x$.

On integrating this Eq. (4.9) taking limit as, from $z = x$ to $z = y$ and taking twice, to take account of the image current with the result then

$$B = 2 \int dB = 2 \times \int \frac{\mu_0 I D dz}{4\pi \left(D^2 + Z^2\right)^{3/2}}$$

$$= \frac{\mu_0 I D}{2\pi} \int \frac{dz}{(D^2 + Z^2)^{3/2}} \tag{4.10}$$

On solving,

$$B = \frac{\mu_0 I D}{2\pi} \cdot \frac{1}{D^2} \left[\frac{Z}{\sqrt{D^2 + Z^2}}\right]_x^y$$

$$= \frac{\mu_0 I}{2\pi D} \left[\frac{y}{(D^2 + y^2)^{1/2}} - \frac{x}{(D^2 + x^2)^{1/2}} \right] \tag{4.11}$$

If the current flow is between a charge center and the ground, i.e., $y = H$ and $x = 0$ then the equation becomes

$$B = \frac{\mu_0 I}{2\pi D} \left[\frac{H}{(H^2 + D^2)^{1/2}} - \frac{O}{(O^2 + D^2)^{1/2}} \right]$$
$$= \frac{\mu_0 I}{2\pi D} \cdot \frac{H}{(H^2 + D^2)^{1/2}} \tag{4.12}$$

Case I:

When $H \gg D$, then the magnetic flux at this point is $B = \frac{\mu_0 I}{2\pi D} \cdot \frac{H}{H} = \frac{\mu_0 I}{2\pi D}$ that is the magnetic field at a very close distance.

Case II:

For the point at a far distance from the discharge, i.e., $H \ll D$, then the magnetic flux at this point is $B = \frac{\mu_0 I H}{2\pi D^2}$. The electric dipole moment of a charge $+ Q$ located at a height H above a conducting plane, earth and its image charge is $M = 2QH$. If the charge $+ Q$ flows to ground, the current flowing equals the rate of change of charge at the source, and thus, $\frac{dM}{dt} = 2 \frac{dQ}{dt} H = 2I H$.

Hence, on substituting, we get, $B = \frac{\mu_0}{4\pi D^2} \cdot \frac{dM}{dt}$ where M is the electric dipole moment.

4.4 Derivation of Jefimenko Equation from Maxwell's Equation

The sources ϱ, volume charge density, and J, volume current density generate electric and magnetic field. The lightning discharge channel produced the electric field and magnetic field, and the Jefimenko equations were derived from Maxwell's equations (Adhikari & Bhandari, 2018a). Let E and B are the electric and magnetic field intensity, μ_0 and ϵ_0 are the permeability and permittivity of the medium. We know the Maxwell's equations are:

(1) $\nabla.E(r, t) = \dfrac{\rho(r', t_r)}{\epsilon_0}$ (2) $\nabla \cdot B(r, t) = 0$

(3) $\nabla \times E(r, t) = -\dfrac{\partial B(r, t)}{\partial t}$

(4) $\nabla \times B(r, t) = \mu_0 J(r', t_r) + \mu_0 \epsilon_0 \dfrac{\partial E(r, t)}{\partial t} = \mu_0 J(r', t_r) + \dfrac{1}{c^2} \dfrac{\partial E(r, t)}{\partial t}$

where c is the speed of light which is the square root of $\frac{1}{\mu_0 \epsilon_0}$.

$E(r, t)$ as the gradient of a scalar potential and $B(r, t)$ remain divergence less, so, we get, $E(r, t) = -\nabla V$ and $B(r, t) = \nabla \times A(r, t)$.

Substituting these two into the Faraday's law, we get,

$$\nabla \times E(r, t) = -\frac{\partial}{\partial t} B(r, t) = -\frac{\partial}{\partial t} (\nabla \times A(r, t))$$

and

$$\nabla \times \left(E(r, t) + \frac{\partial A(r, t)}{\partial t} \right) = 0.$$

Here, the term whose curl gives zero, therefore it is the gradient of a scalar.

$$\left(E(r, t) + \frac{\partial A(r, t)}{\partial t} \right) = -\nabla V \quad \Rightarrow E(r, t) = -\nabla V - \frac{\partial A(r, t)}{\partial t}$$

or,

$$\nabla \cdot E(r, t) = \frac{1}{\epsilon_0} \rho(r', t_r) \Rightarrow \nabla \cdot \left(-\nabla V - \frac{\partial A(r, t)}{\partial t} \right) = \frac{1}{\epsilon_0} \rho(r', t_r) \qquad (4.13)$$

Again, from Maxwell fourth relation,

$$\nabla \times B(r, t) = \mu_0 J(r', t_r) + \mu_0 \epsilon_0 \frac{\partial E(r, t)}{\partial t}.$$

or,

$$\nabla \times (\nabla \times A(r, t)) = \mu_0 J(r', t_r) + \mu_0 \epsilon_0 \cdot \frac{\partial}{\partial t} \left(-\nabla V - \frac{\partial A(r, t)}{\partial t} \right).$$

or,

$$\nabla(\nabla \cdot A(r, t)) - \nabla^2 A(r, t) = \mu_0 J(r', t_r) - \mu_0 \epsilon_0 \nabla \left(\frac{\partial V}{\partial t} \right) - \mu_0 \epsilon_0 \frac{\partial^2 A(r, t)}{\partial t^2}.$$

or,

$$\nabla^2 A(r, t) - \mu_0 \epsilon_0 \frac{\partial^2 A(r, t)}{\partial t^2} - \nabla \left(\nabla \cdot A(r, t) + \mu_0 \epsilon_0 \frac{\partial V}{\partial t} \right) = -\mu_0 J(r', t_r).$$

$$(4.14)$$

These two Eqs. (4.13) and (4.14) contain all the information in Maxwell's equations.

In static case $\nabla^2 V = -\frac{\rho(r', t_r)}{\epsilon_0}$ and $\nabla^2 A(r, t) = -\mu_0 J(r', t_r)$.

The solutions of these equations are,

$$V(r) = \frac{1}{4\pi \, \epsilon_0} \int \frac{\rho(r')}{r} dt$$

and

$$A(r) = \frac{\mu_0}{4\pi} \int \frac{J(r')}{r} dt.$$

For non-static sources the scalar potential is

$$V(r, t) = \frac{1}{4\pi \, \epsilon_0} \int \frac{\rho(r', t_r)}{r} dt.$$

and the vector potential is

$$A(r, t) = \frac{\mu_0}{4\pi} \int \frac{J(r', t_r)}{r} dt.$$

where $\varrho(r', t)$ is the charge density that prevailed at point r' at the retarded time t_r, where $t_r = t - r/c$. The relation between the two potential functions (scalar and vector potentials) is given by the Lorentz condition,

$$\nabla \cdot A(r, t) + \frac{1}{c^2} \frac{\partial v}{\partial t} = 0.$$

or,

$$\nabla \cdot A(r, t) + \mu_0 \, \epsilon_0 \frac{\partial v}{\partial t} = 0.$$

For any scalar function, we can do any operation, due to this, there is no affected on the electric and magnetic field (E and B). Such changes in scalar potential (A, V) are called Gauge transformation. Adhikari and Bhandari (2018b) described and simplified that and calculated the Laplacian of $V(r, t)$, by taking gradient, then,

$$
\begin{aligned}
\nabla V(r, t) &= \frac{1}{4\pi \epsilon_0} \int \left[\nabla \left(\frac{\rho(r', t_r)}{r} \right) d\tau \right] \\
&= \frac{1}{4\pi \epsilon_0} \int \left[(\nabla \rho(r', t_r)) \frac{1}{r} + \rho(r', t_r) \nabla \left(\frac{1}{r} \right) \right] d\tau \\
&= \frac{1}{4\pi \epsilon_0} \int \left[-\frac{\dot{\rho}(r', t_r) \, \hat{r}}{c \quad r} - \rho(r', t_r) \frac{\hat{r}}{r^2} \right] d\tau
\end{aligned}
$$

Again, taking the divergence of $\nabla V(r,t) = \nabla.\nabla V(r,t)$

$$\nabla^2 V = \frac{1}{4\pi \epsilon_0} \int \left\{ \begin{array}{l} -\frac{1}{c}\left[\frac{\hat{r}}{r}.(\nabla\dot{\rho}(r',t_r)) + \dot{\rho}(r',t_r)\nabla.\frac{r}{r}\right] \\ -\left[\frac{r}{r^2}.\rho(r',t_r)\nabla + \nabla\rho(r',t_r).\left(\frac{r}{r^2}\right)\right] \end{array} \right\} dt$$

$$= \frac{1}{4\pi \epsilon_0} \int \left[\frac{1}{c^2}\frac{\ddot{\rho}(r',t_r)}{r} - 4\pi\rho.\delta^3(r)\right]dt$$

$$= \frac{1}{c^2}\frac{\partial^2 V(r,t)}{\partial t^2} - \frac{1}{\epsilon_0}\rho(r',t_r)$$

[By substituting $\nabla\dot{\rho}(r',t_r) = -\frac{1}{c}\ddot{\rho}(r',t_r)\nabla r$, and $\nabla.\frac{r}{r} = \frac{1}{r^2}$, $\nabla.\frac{r}{r^2} = 4\Pi\delta^3(r)$].
This is three-dimensional Dirac delta function.
The time derivative of scalar potential A is $\frac{\partial A(r,t)}{\partial t} = \frac{\mu_0}{4\pi}\int\frac{J(r',t_r)}{r}d\tau$ and using the relation $c^2 = \frac{1}{\mu_0\epsilon_0}$. Putting both of them together in the equation, we get,

$$E(r,t) = -\nabla V(r,t) - \frac{\partial A(r,t)}{\partial t}$$

$$= -\left[-\frac{1}{4\pi \epsilon_0}\int\left\{\frac{\dot{\rho}(r',t_r)}{cr}r + \frac{\rho(r',t_r)}{r^2}r\right\}d\tau\right] - \frac{\mu_0}{4\pi}\int\frac{J(r',t_r)}{r}d\tau$$

Then we get,

$$E(r,t) = \frac{1}{4\pi \epsilon_0}\int\left[\frac{\rho(r',t_r)}{r^2}r + \frac{\dot{\rho}(r',t_r)}{cr}r - \frac{\dot{J}(r',t_r)}{c^2 r}\right]d\tau.$$

This is the time-dependent generalization of coulomb's law.
For the magnetic field (B), the curl of A contains two terms:

$$\nabla \times A(r,t) = \frac{\mu_0}{4\pi}\int\nabla\times\left(\frac{J(r',t_r)}{r}\right)d\tau$$

$$= \frac{\mu_0}{4\pi}\int\left[\frac{1}{r}\nabla\times J(r',t_r) - J(r',t_r)\times\nabla\left(\frac{1}{r}\right)\right]dt$$

$$= \frac{\mu_0}{4\pi}\int\left[\frac{J(r',t_r)}{r^2} + \frac{\dot{J}(r',t_r)}{c.r}\right]\times r d\tau$$

Hence, $B(r,t) = \frac{\mu_0}{4\pi}\int\left[\frac{J(r',t_r)}{r^2} + \frac{\dot{J}(r',t_r)}{c.r}\right]\times r d\tau.$
This is the time-dependent generalization of Biot–Savart law.

4.5 Expressions for Electric and Magnetic Fields

Maxwell formulated the electric field and magnetic field in terms of charge and current densities and hence found the relation between them. The limit of an extended charge be a point charge when the size goes to zero. As shown in Fig. 4.3, the lightning channel starts from a point on the ground and it moves toward the cloud as a return stroke (Adhikari, 2019a). Griffith (1999) explained about the relation between the electric and magnetic fields. When the charge moves with the velocity "v" at the retarded time, then the charge should be reduced by the retardation factor of $\left[1 - r \cdot \frac{v}{c}\right]^{-1}$.

Then we have the relation,

$$\int \rho(r', t_r) d\tau = \frac{q}{\left(1 - r \cdot \frac{v}{c}\right)}.$$

Then, on substituting, we get the potential,

$$V(r, t) = \frac{1}{4\pi \, \epsilon_0} \cdot \int \frac{\rho(r', t_r)}{r} d\tau.$$

Hence gives potential $V(r, t) = \frac{1}{4\pi \, \epsilon_0} \frac{qc}{(rc - r.v)}$ where r is the vector from the retarded position to the field point $\boldsymbol{r}$.

Again,

$$A(r, t) = \frac{\mu_0}{4\pi} \int \frac{\rho(r', t_r) v(t_r)}{r} \, dtau$$

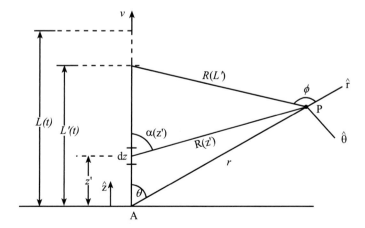

Fig. 4.3 Geometry of problem in the treatment of retardation effects (Cooray, 2003)

$$= \frac{\mu_0}{4\pi} \frac{v}{r} \int \rho(r', t_r) \, d\tau = \frac{\mu_0}{4\pi} \frac{qcv}{(rc - r.v)} = \frac{v}{c^2} V(r, t).$$

Gradient of

$$V = \nabla V = \nabla \left\{ \frac{1}{4\pi \, \epsilon_0} \frac{qc}{(rc - r.v)} \right\} = \frac{qc}{4\pi \, \epsilon_0} \cdot \frac{-1}{(rc - r.v)^2} . \nabla (rc - r.v).$$

Since $r = c(t - t_r)$ and $t_r = t - \frac{r}{c}$, $\nabla r = -c.\nabla t_r$

$$\nabla(r.v) = (r.\nabla)v + (v.\nabla)r + r \times \nabla \times v + v \times (\nabla \times r)$$

Evaluating these terms (one by one)

$$(r.\nabla)v = \left(r_x \frac{\partial}{\partial x} + r_y \frac{\partial}{\partial y} + r_z \frac{\partial}{\partial z} \right) v(t_r)$$

$$= r_x . \frac{\partial v}{\partial t_r} . \frac{\partial t_r}{\partial x} + r_y . \frac{\partial v}{\partial t_r} \frac{\partial t_r}{\partial y} + r_z . \frac{\partial v}{\partial t_r} \frac{\partial t_r}{\partial z}$$

$$= \frac{\partial v}{\partial t_r} \left(r_x \frac{\partial t_r}{\partial x} + r_y \frac{\partial t_r}{\partial y} + r_z \frac{\partial t_r}{\partial z} \right)$$

$$= \frac{\partial v}{\partial t_r} \left(r_x \frac{\partial}{\partial x} + r_y \frac{\partial}{\partial y} + r_z \frac{\partial}{\partial z} \right) t_r = a(r.\nabla t_r)$$

where $a \equiv \dot{v}$ is the acceleration of the particle at the retarded time.

$$(v.\nabla)r = (v.\nabla)r - (v.\nabla)w \quad [\because r = r - w]$$

and

$$(v.\nabla)r = \left(v_x \frac{\partial}{\partial x} + v_y \frac{\partial}{\partial y} + v_z \frac{\partial}{\partial z} \right) (Xx + Yy + Zz)$$

$$= (V_x x + V_y y + V_z z) = v$$

While

$$(v.\nabla)w = \left(v_x \frac{\partial}{\partial x} + v_y \frac{\partial}{\partial y} + v_z \frac{\partial}{\partial z} \right) w.(t_r)$$

$$= v_x . \frac{\partial w}{\partial t_r} . \frac{\partial t_r}{\partial x} + v_y \frac{\partial w}{\partial t_r} \frac{\partial t_r}{\partial y} + v_z \frac{\partial w}{\partial t_r} \frac{\partial t_r}{\partial z}$$

$$= \frac{\partial w}{\partial t_r} \left(v_x \frac{\partial t_r}{\partial x} + v_y \frac{\partial t_r}{\partial y} + v_z \frac{\partial t_r}{\partial z} \right) = v(v.\nabla t_r)$$

Hence, $(v.\nabla)r = v - v(v.\nabla t_r)$.

For third term, $\nabla \times v = -a \times \nabla t_r$.

Similarly, $\nabla \times w = -v \times \nabla t_r$.
Again,

$$\nabla \times r = \nabla \times (r - w) = \nabla \times r - \nabla \times w$$
$$= 0 - \nabla \times w = -v \times \nabla t_r$$

Substituting these all values then we get,

$$\nabla(r.v) = a(r.\nabla t_r) + v - v(v.\nabla t_r) - r \times a \times \nabla t_r + v \times v \times \nabla t_r$$
$$= v + (r.a - v^2)\nabla t_r$$

Hence,

$$\nabla \vec{v} = \frac{qc}{4\pi \epsilon_0} \cdot \frac{1}{(rc - r \cdot v)^2}\left[v + (c^2 - v^2 + r \cdot a)\nabla t_r\right].$$

Now, for ∇t_r, $\nabla r = -c\nabla t_r$.
So,

$$-c\nabla t_r = \nabla r = \nabla\sqrt{r, r} = \frac{1}{2}\frac{1}{\sqrt{r.r}}\nabla(r.r)$$
$$= \frac{1}{r}[(r.\nabla)r + r \times (\nabla \times r)]$$

But $(r.\nabla)r = r - v(r.\nabla t_r)$.
Same as above, $\nabla \times r = v \times \nabla t_r$.
Thus,

$$-c\nabla t_r = \frac{1}{r}[r - v(r.\nabla t_r) + r \times (v \times \nabla t_r)] = \frac{1}{r}[r - (r.v)\nabla t_r)]$$
$$\therefore \quad \nabla t_r = \frac{-r}{rc - r.v}$$

Hence, the gradient of

$$V = \Delta V = \frac{1}{4\pi \epsilon_0}\frac{qc}{(rc - r.v)^3}\left[(rc - r.v)v - (c^2 - v^2) + (r.a)r\right]$$

Similarly, for

$$\frac{\partial A}{\partial t} = \frac{1}{4\pi \epsilon_0}\frac{qc}{(rc - \hat{r}.v)^3}\left[(rc - r.v)\left(-v + \frac{ra}{c}\right) + \frac{r}{c}\{c^2 - v^2 + (r.a)\}v\right].$$

Then $u \equiv cr - v$.
Then,

$$E(r, t) = \frac{q}{4\pi \, \epsilon_0} \frac{r}{(r - u)^3} \left[(c^2 - v^2)u + r \times (u \times a) \right].$$

Meanwhile,

$$\nabla \times A(r, t) = \frac{1}{c^2} \nabla \times (vv) = \frac{1}{c^2} [v(\nabla \times v) - v \times \nabla v].$$

or,

$$\nabla \times A(r, t) = \frac{1}{c} \frac{q}{4\pi \, \epsilon_0} \frac{1}{(u.r)^3} r \times \left[(c^2 - v^2)v + (r.a)v + (r.u)a \right].$$

Hence, $B(r, t) = \frac{1}{c} r \times E(r, t)$.

4.6 Time-Varying Fields from Lightning: Treatment of Retardation Effects

In the lightning discharges channel comparing the overall length from the ground to cloud, the thickness of it should be very thin. The flow of charges in the channel with the high speed as in the order of the velocity of light, the retardation effect should be considered to calculate the electric and magnetic fields. For this it is modeled that a linear antenna has some line charge density distribution or current distribution that changes with time. Therefore, to calculate electric and magnetic fields from lightning, the analytical expressions for electric field and magnetic field vary with time of the source distribution of lightning discharges are presented. The speed of light is maximum which is in magnitude of 3×10^8 m/s, and the velocity of lightning return stroke wave-front is only from thirty to fifty percent of the speed of the light. Even the speed is 30–50% of that of light, the velocity of the stroke is also very high. Hence, the time taken to travel the light from the return stroke channel to the field measurement point cannot be ignored. The observer observes the current on the channel of return stroke at the field measuring point at the time earlier and cannot observe the true length of the channel. Thottappillil et al (1998), Thottappillil et al. (2001), Cooray (2003), Adhikari (2018) described about the retarded channel lengths are to be used for the calculation of the electric and magnetic fields due to lightning return stroke channel.

Let $L(t)$ be the actual length of the lightning return stroke wave-front which moves with velocity v and the apparent length of the return stroke channel is $L'(t)$ from the fixed point A as shown in Fig. 4.3. Let P be the point of observer, which observes the channel emerging from the point A at the time r/c. There is some difference between the two real and apparent lengths (also called retarded length). From the simple definition, the time taken is the ratio of distances to the velocity. But here the time required for the return stroke wave-front to a point P is the sum of time from the point A to travel the distances $L'(t)$ and the time to travel the distance $R(L')$. Hence, the total time is,

$$t = \frac{L'(t)}{v} + \frac{R(L')}{c},$$

where $R(L') = \sqrt{r^2 + L'^2 - 2L'r \cos \theta}$ from the cosine law and θ be the angle between r and L'. Then

$$t = \frac{L'(t)}{v} + \frac{\sqrt{r^2 + L'^2(t) - 2L'(t)r \cos \theta}}{c}.$$

or,

$$\left(t - \frac{L'(t)}{v}\right)^2 = \frac{r^2 + L'^2(t) - 2rL'(t) \cos \theta}{c^2}.$$

or,

$$c^2 \left(vt - L'(t)\right)^2 = v^2 \left[r^2 + L'^2(t) - 2rL'(t) \cos \theta\right].$$

or,

$$c^2 \left[v^2 t^2 + L'^2(t) - 2L'(t)vt\right] = r^2 v^2 + v^2 L'^2(t) - 2rL'(t)v^2 \cos \theta.$$

or,

$$c^2 L'^2(t) + c^2 v^2 t^2 - 2L'(t)vtc^2 = r^2 v^2 + L'^2(t)v^2 - 2rL'(t)v^2 \cos \theta.$$

or,

$$L'^2(t)\left(c^2 - v^2\right) - 2L'(t)vtc^2 + 2L'(t)rv^2 \cos \theta = r^2 v^2 - c^2 v^2 t^2.$$

or,

$$L'^2(t) - \frac{2.L'(t)\left(vtc^2 - rv^2 \cos \theta\right)}{c^2 - v^2} = \frac{r^2 v^2 - c^2 v^2 t^2}{c^2 - v^2}.$$

or,

$$\left[L'(t) - \left(\frac{vtc^2 - rv^2 \cos \theta}{c^2 - v^2}\right)\right]^2 = \frac{r^2 v^2 - c^2 v^2 t^2}{c^2 - v^2} + \left(\frac{vtc^2 - rv^2 \cos \theta}{c^2 - v^2}\right)^2.$$

or,

$$\left[L'(t) - \left\{\frac{vt - r\cos\theta\left(\frac{v^2}{c^2}\right)}{1 - \frac{v^2}{c^2}}\right\}\right]^2$$

$$= \frac{r^2\frac{v^2}{c^2} - v^2t^2}{1 - \frac{v^2}{c^2}} + \left(\frac{vt - r\cos\theta\frac{v^2}{c^2}}{1 - \frac{v^2}{c^2}}\right)^2$$

$$= \frac{\left(1 - \frac{v^2}{c^2}\right)\left(r^2\frac{v^2}{c^2} - v^2t^2\right) + v^2t^2 + r^2\cos^2\theta\left(\frac{v^2}{c^2}\right)^2 - 2v.t.r\cos\theta\frac{v^2}{c^2}}{\left(1 - \frac{v^2}{c^2}\right)}.$$

Taking numerator only

$$\left(1 - \frac{v^2}{c^2}\right)\left(r^2\frac{v^2}{c^2} - v^2t^2\right) + v^2t^2 + r^2\cos^2\theta\left(\frac{v^2}{c^2}\right)^2 - 2vtr\cos\theta\frac{v^2}{c^2}$$

$$= \frac{r^2v^2}{c^2} - r^2\left(\frac{v^2}{c^2}\right)^2 - v^2t^2 + v^2t^2.\frac{v^2}{c^2} + v^2t^2 + r^2\cos^2\theta\left(\frac{v^2}{c^2}\right)^2 - 2vtr\cos\theta\frac{v^2}{c^2}$$

$$= r^2\frac{v^2}{c^2} - r^2\left(\frac{v^2}{c^2}\right)^2 + r^2\cos^2\theta\left(\frac{v^2}{c^2}\right)^2 + \frac{v^2}{c^2}\left(v^2t^2 - 2vt\cos\theta\right)$$

$$= r^2\frac{v^2}{c^2}\left(1 - \frac{v^2}{c^2}\right) + \frac{r^2v^2}{c^2}\left[\frac{v^2}{c^2}\cos^2\theta + \frac{v^2t^2 - 2vtr\cos\theta}{r^2}\right]$$

$$= r^2\frac{v^2}{c^2}\left(1 - \frac{v^2}{c^2}\right) + \frac{r^2v^2}{c^2}\left[\frac{v^2t^2}{c^2} - \frac{2vt\cos\theta}{r} + \frac{v^2}{c^2}\cos^2\theta\right]$$

$$= \left[r^2\frac{v^2}{c^2}\left(1 - \frac{v^2}{c^2}\right) + \frac{v^2t^2}{r^2} - \frac{2vt\cos\theta}{r} + \frac{v^2}{c^2}\cos^2\theta\right]$$

Now,

$$\left\{L'(t) - \left(\frac{vt - r\cos\theta\frac{v^2}{c^2}}{1 - \frac{v^2}{c^2}}\right)\right\}^2 = \frac{\frac{r^2v^2}{c^2}\left[\left(1 - \frac{v^2}{c^2}\right) + \frac{v^2t^2}{r^2} - \frac{2vt\cos\theta}{r} + \frac{v^2}{c^2}\cos^2\theta\right]}{\left(1 - \frac{v^2}{c^2}\right)^2}.$$

Taking square root on both sides

$$L'(t) - \left(\frac{vt - r\cos\theta\frac{v^2}{c^2}}{1 - \frac{v^2}{c^2}}\right) = \pm\frac{rv/c}{1 - v^2/c^2}\left[\left(1 - \frac{v^2}{c^2}\right) + \frac{v^2t^2}{r^2} - \frac{2vt\cos\theta}{r} + \frac{v^2}{c^2}\cos^2\theta\right]^{1/2}$$

$$L'(t) = \frac{vt - r\cos\theta v^2/c^2}{1 - v^2/c^2} \pm \frac{rv/c}{1 - v^2/c^2}\left[\left(1 - \frac{v^2}{c^2}\right) + \frac{v^2t^2}{r^2} - \frac{2vt\cos\theta}{r} + \frac{v^2}{c^2}\cos^2\theta\right]^{1/2}$$

$$L'(t) = \frac{r}{1 - v^2/c^2}\left\{-\cos\theta\frac{v^2}{c^2} + \frac{vt}{r} \pm \frac{v}{c}\sqrt{\left(1 - \frac{v^2}{c^2}\right) + \frac{v^2t^2}{r^2} + \frac{v^2}{c^2}\cos^2\theta - \frac{2vt}{r}\cos\theta}\right\}$$

1. If the ground is treated as being perfectly conducting, then the angle θ be replaced by $(180 - \theta)$. Then we get,

$$L''(t) = \frac{r}{1 - v^2/c^2} \left\{ \frac{v^2}{c^2} \cos\theta + \frac{vt}{r} \pm \frac{v}{c} \sqrt{\left(1 - \frac{v^2}{c^2}\right) + \frac{v^2 t^2}{r^2} + \frac{v^2}{c^2} \cos^2\theta + \frac{2vt}{r} \cos\theta} \right\}$$

2. If all the channel sections were equidistant from the observer at a distance r, the discharge should be in circular arc. Then $\theta = 0$, so,

$$L'(t) = \frac{r}{1 - v^2/c^2} \left\{ \frac{-v^2}{c^2} + \frac{vt}{r} \pm \frac{v}{c} \sqrt{\left(1 - \frac{v^2}{c^2}\right) + \frac{v^2 t^2}{r^2} + \frac{v^2}{c^2} - \frac{2vt}{r}} \right\}$$

$$= \frac{r}{1 - v^2/c^2} \left\{ \frac{-v^2}{c^2} + \frac{vt}{r} \pm \frac{v}{c} \sqrt{1 - \frac{2vt}{r} + \frac{v^2 t^2}{r^2}} \right\}$$

$$= \frac{r}{1 - v^2/c^2} \left\{ \frac{-v^2}{c^2} + \frac{vt}{r} \pm \frac{v}{c} \left(1 - \frac{vt}{r}\right) \right\}$$

Taking negative sign,

$$L'(t) = \frac{r}{1 - v^2/c^2} \left\{ \frac{-v^2}{c^2} + \frac{vt}{r} - \frac{v}{c} \left(1 - \frac{vt}{r}\right) \right\}$$

$$= \frac{r}{1 - v^2/c^2} \left\{ -\frac{v}{c} - \frac{v^2}{c^2} + \frac{vt}{r} \left(1 + \frac{v}{c}\right) \right\}$$

$$= \frac{r}{1 - v^2/c^2} \left\{ -\frac{v}{c} \left(1 + \frac{v}{c}\right) + \frac{vt}{r} \left(1 + \frac{v}{c}\right) \right\}$$

$$= \frac{r}{1 - v^2/c^2} \left\{ \left(-\frac{v}{c} + \frac{vt}{r}\right) \left(1 + \frac{v}{c}\right) \right\}$$

$$= \frac{r}{1 - v/c} \left(-\frac{v}{c} + \frac{vt}{r}\right) = \frac{v}{1 - v/c} \left(t - \frac{r}{c}\right)$$

Taking positive sign

$$L'(t) = \frac{r}{1 - v^2/c^2} \left\{ -\frac{v^2}{c^2} + \frac{vt}{r} + \frac{v}{c} \left(1 - \frac{vt}{r}\right) \right\}$$

$$= \frac{r}{1 - v^2/c^2} \left\{ \frac{v}{c} \left(1 - \frac{v}{c}\right) + \frac{vt}{r} \left(1 - \frac{v}{c}\right) \right\}$$

$$= \frac{r}{1 - v^2/c^2} \left\{ \left(\frac{v}{c} + \frac{vt}{r}\right) \left(1 - \frac{v}{c}\right) \right\}$$

$$= \frac{r}{1 + v/c} \left(\frac{v}{c} + \frac{vt}{r}\right) = \frac{v}{1 + v/c} \left(t + \frac{r}{c}\right)$$

Taking negative sign, then we get

$$L'(t) = \frac{v}{1 - v/c}\left(t - \frac{r}{c}\right)$$

Taking positive sign, then we get,

$$L'(t) = \frac{v}{1 + v/c}\left(t + \frac{r}{c}\right)$$

The same condition is applied for image channel length $L''(t)$ then we get

$$L''(t) = \frac{r}{1 - v^2/c^2}\left\{\frac{v^2}{c^2} + \frac{vt}{r} \pm \frac{v}{c}\sqrt{\left(1 - \frac{v^2}{c^2}\right) + \frac{v^2t^2}{r^2} + \frac{v^2}{c^2} + \frac{2vt}{r}}\right\}$$

$$L''(t) = \frac{r}{1 - v^2/c^2}\left\{\frac{v^2}{c^2} + \frac{vt}{r} \pm \frac{v}{c}\sqrt{\left(1 + \frac{vt}{r}\right)^2}\right\}$$

$$= \frac{r}{1 - v^2/c^2}\left\{\frac{v^2}{c^2} + \frac{vt}{r} \pm \frac{v}{c}\left(1 + \frac{vt}{r}\right)\right\}$$

Taking positive sign

$$L''(t) = \frac{r}{1 - v^2/c^2}\left\{\frac{v^2}{c^2} + \frac{vt}{r} + \frac{v}{c}\left(1 + \frac{vt}{r}\right)\right\}$$

$$= \frac{r}{1 - v^2/c^2}\left\{\frac{v^2}{c^2} + \frac{vt}{r} + \frac{v}{c} + \frac{v^2t}{cr}\right\}$$

$$= \frac{r}{1 - v^2/c^2}\left\{\frac{v}{c}\left(1 + \frac{v}{c}\right) + \frac{vt}{r}\left(1 + \frac{v}{c}\right)\right\}$$

$$= \frac{r}{1 - v/c}\left(\frac{v}{c} + \frac{vt}{r}\right) = \frac{v}{1 - v/c}\left(t + \frac{r}{c}\right)$$

Taking negative sign,

$$L''(t) = \frac{r}{1 - v^2/c^2}\left\{\frac{v^2}{c^2} + \frac{vt}{r} - \frac{v}{c}\left(1 + \frac{vt}{r}\right)\right\}$$

$$= \frac{r}{1 - v^2/c^2}\left\{\frac{v^2}{c^2} + \frac{vt}{r} - \frac{v}{c} - \frac{v^2t}{cr}\right\}$$

$$= \frac{r}{1 - v^2/c^2}\left\{-\frac{v}{c}\left(1 - \frac{v}{c}\right) + \frac{vt}{r}\left(1 - \frac{v}{c}\right)\right\}$$

$$= \frac{r}{1 - v^2/c^2}\left\{\left(-\frac{v}{c} + \frac{vt}{r}\right)\left(1 - \frac{v}{c}\right)\right\}$$

$$= \frac{r}{1 + v/c}\left(-\frac{v}{c} + \frac{vt}{r}\right) = \frac{v}{1 + v/c}\left(t - \frac{r}{c}\right)$$

Note: Hence for retarded time, we take $\left(t - \frac{r}{c}\right)$ so, for both apparent channel length $L'(t) = \frac{v}{1-v/c}\left(t - \frac{r}{c}\right)$ and image channel length $L''(t) = \frac{v}{1+v/c}\left(t - \frac{r}{c}\right)$ obtained by taking negative sign. Similarly, for advanced time when we take positive sign, then we get $L'(t) = \frac{v}{1+v/c}\left(t + \frac{r}{c}\right)$ and $L''(t) = \frac{v}{1-v/c}\left(t + \frac{r}{c}\right)$.

Hence, apparent channel length $L'(t)$ for retarded potential $= \frac{v}{1-v/c}\left(t - \frac{r}{c}\right)$ and image channel length $L''(t)$ for retarded potential $= \frac{v}{1+v/c}\left(t - \frac{r}{c}\right)$.

4.7 Determination of Electric Field in Terms of Current and Charge by the Continuity Equation Approach

To determine an expression for the electric field in terms of current density and charge density, the continuity equation approach is used. By using scalar and vector potential, it relates the charge density and current density. Adhikari (2019b) explains about the retardation time for the return stroke lightning channel. The relation between the charge density and current density at the retarded time is

$$\frac{\partial \rho\left(z', t_r\right)}{\partial t} = -\left.\frac{\partial i\left(z', t_r\right)}{\partial z'}\right|_{t_r = \text{constant}}.$$

where t_r is the retarded time which is equal to $t_r = t - \frac{R(z')}{c}$.

As shown in Fig. 4.3, $L(t)$ be the actual length, $L'(t)$ be the apparent length of the lightning return stroke wave-front which moves with velocity v. The partial derivative of the current in the right-hand side of the above equation, with source coordinate (z') is balanced with the rate of change of charge density. Let us consider, the return stroke starting from the point A on the ground in which $z' = 0$, then the charge is

$$Q\left(t - \frac{r}{c}\right) = -\int_{r/c}^{t} i\left(0, \tilde{t} - \frac{r}{c}\right)d\tau$$

The scalar potential due to the whole lightning channel is,

$$\phi(r, t) = \frac{1}{4\pi \epsilon_0}\frac{Q(t - r/c)}{r} + \frac{1}{4\pi \epsilon_0}\int_{0}^{L'(t)} \rho\left(z', t - \frac{R(z')}{c}\right)dz'$$

The negative gradient of the scalar potential by using the spherical coordinate system is given by

$$-\nabla\phi = \frac{1}{4\pi\,\epsilon_0}\left[\begin{array}{c}\hat{r}\dfrac{\partial}{\partial r}\dfrac{Q(t-r/c)}{r}+\hat{r}\dfrac{\partial}{\partial r}\displaystyle\int_{0}^{L'(t)}\dfrac{1}{R(z')}\rho\!\left(z',\,t-\dfrac{R(z')}{c}\right)dz'\\[4mm]+\hat{\theta}\dfrac{1}{r}\displaystyle\int_{0}^{L'(t)}\dfrac{\partial}{\partial\theta}\dfrac{\rho\!\left(z',t-\dfrac{R(z')}{c}\right)}{R(z')}dz'\end{array}\right]$$

From the Fig. 4.3, we can write,

$$R^2\!\left(z'\right)=r^2+z'^2-2rz'\cos\theta$$

or,

$$R\!\left(z'\right)=\sqrt{r^2+z'^2-2rz'\cos\theta}.$$

Here, the $R(z')$ is the function of both r and θ, so, on the partial differentiation,

$$\frac{\partial R\!\left(z'\right)}{\partial\theta}=\frac{1}{2}(r^2+z'^2-2rz'\cos\theta)^{-1/2}.2rz'\sin\theta$$

$$\frac{\partial R\!\left(z'\right)}{\partial\theta}=\frac{rz'\sin\theta}{R(z')}$$

Again,

$$\frac{\partial R\!\left(z'\right)}{\partial r}=\frac{1}{2}(r^2+z'^2-2rz'\cos\theta)^{-1/2}.(2r-2z'\cos\theta)=\frac{r-z'\cos\theta}{R(z')}$$

Now,

$$\frac{\partial\rho}{\partial r}=\frac{\partial\rho}{\partial R}.\frac{\partial R}{\partial r}=-\frac{1}{c}\frac{\partial\rho}{\partial t}.\frac{r-z'\cos\theta}{R(z')}.$$

and

$$\frac{\partial\rho}{\partial\theta}=\frac{\partial\rho}{\partial R}.\frac{\partial R}{\partial\theta}=-\frac{1}{c}\frac{\partial\rho}{\partial t}.\frac{rz'\sin\theta}{R(z')}.$$

where $\rho=\rho\!\left(z',t-\dfrac{R(z')}{c}\right)$

$$\frac{\partial\rho}{\partial t}=\frac{\partial\rho}{\partial z'}but\frac{\partial\rho}{\partial R}=-\frac{1}{c}\frac{\partial\rho}{\partial t}$$

On substituting all these values in the above equation

$$-\nabla\phi = \frac{1}{4\pi\,\epsilon_0}\left[\hat{r}\,\frac{\partial}{\partial r}\,\frac{Q(t-r/c)}{r} + \hat{r}\,\frac{\partial}{\partial r}\int_0^{L'(t)}\frac{1}{R(z')}\rho\left(z',t-\frac{R(z')}{c}\right)dz'\right.$$

$$\left. -\hat{\theta}\,\frac{1}{r}\int_0^{L'(t)}\frac{\partial}{\partial\theta}\,\frac{\rho\left(z',t-\frac{R(z')}{c}\right)}{R(z')}dz'\right]$$

or,

$$-4\pi\,\epsilon_0\,\nabla\phi = -\hat{r}\,\frac{\partial}{\partial r}\,\frac{Q\left(t-\frac{r}{c}\right)}{r}$$

$$-\hat{r}\int_0^{L'(t)}\left[\frac{d\rho\left(z',t-\frac{R(z')}{c}\right)}{\partial r}\cdot\frac{1}{R(z')} + \rho\left(z',t-\frac{R(z')}{c}\right) - 4\pi\,\epsilon_0\,\nabla\phi\right]$$

$$= -\hat{r}\,\frac{\partial}{\partial r}\,\frac{Q\left(t-\frac{r}{c}\right)}{r}$$

$$-\hat{r}\int_0^{L'(t)}\left[\frac{d\rho\left(z',t-\frac{R(z')}{c}\right)}{\partial r}\cdot\frac{1}{R(z')} + \rho\left(z',t-\frac{R(z')}{c}\right) - \frac{1}{R^2(z')}\cdot\frac{\partial R}{\partial r}\right]dz'$$

$$-\frac{\hat{\theta}}{r}\int_0^{L'(t)}\left[\frac{\partial\rho\left(z',t-\frac{R(z')}{c}\right)}{d\theta}\,\frac{1}{R(z')} + \left(-\frac{1}{R^2(z')}\right)\frac{\partial R}{\partial\theta}\cdot\rho\right]dz$$

$$-\left[\frac{\hat{r}\rho(L')}{rR(L')}\,\frac{dL'}{dr}\right] - \left[\frac{\hat{\theta}\rho(L')}{rR(L')}\,\frac{dL'}{d\theta}\right]$$

or,

$$-4\pi\,\epsilon_0\,\nabla\phi = \hat{r}\int_0^{L'(t)}\left[\frac{r-z'\cos\theta}{R^3(z)}\,\rho + \frac{r-z'\cos\theta}{cR^2(z')}\,\frac{\partial\rho}{\partial t}\right]dz'$$

$$+\hat{\theta}\int_0^{L'(t)}\left[\frac{z'\sin\theta}{R^3(z)}\,\rho + \frac{z'\sin\theta}{cR^2(z')}\,\frac{\partial\rho}{\partial t}\right]$$

$$\times\hat{\theta}\int_0^{L'(t)}\left[\frac{z'\sin\theta}{R^3(z)}\,\rho + \frac{z'\sin\theta}{cR^2(z')}\,\frac{\partial\rho}{\partial t}\right]dz'$$

$$-1\hat{r}\,\frac{\partial(Q/r)}{\partial r} - \hat{r}\,\frac{\rho(L)}{R(L)}\,\frac{\partial L'}{\partial r} - \hat{\theta}\,\frac{\rho(L')}{rR(L')}\,\frac{\partial L'}{\partial\theta}$$

Here, $\rho \equiv \rho\left(z',t-\frac{R(z')}{c}\right)$

$$\rho(L') = \rho\left(L', t - \frac{R(L')}{c}\right)$$

The vector potential at P due to the whole retarded length $L'(t)$ is

$$A(r, t) = \frac{\mu_0}{4\pi} \int\limits_0^{L'(t)} \frac{i\left(z', t\frac{R(z')}{c}\right)}{R(z')} dz' \hat{z}$$

Now, the negative rate of change of vector potential is

$$-\frac{\partial A}{\partial t} = \frac{1}{4\pi \epsilon_0}(-\hat{z}) \int\limits_0^{L'(t)} \left[\frac{1}{c^2 R(z')} \frac{\partial i\left(z', t - \frac{R(z')}{c}\right)}{\partial t} dz' - \hat{z} \frac{i\left(L', t - \frac{R(L')}{c}\right)}{c^2 R(L')} \frac{dL'}{dt} \right]$$

where $\hat{z} = \hat{r} \cos\theta - \hat{\theta} \sin\theta$.

On combining these two equations, then we get the total electric field at the point P.

$$E(\rho, \theta, \tau) = -\nabla\phi + \frac{\partial A}{\partial t}$$

$\therefore E(\rho, \theta, \tau)$

$$= \frac{1}{4\pi \epsilon_0} \left[\hat{r} \int\limits_0^{L'(t)} \left\{ \begin{array}{c} \frac{r - z' \cos\theta}{R^3(z')} \rho\left(z', t - \frac{R(z')}{c}\right) \\ + \frac{r - z' \cos\theta}{cR^2(z')} \frac{\partial \rho(z', t - R(z')/c)}{\partial t} \end{array} \right\} dz' \right]$$

$$+ \hat{\theta} \int\limits_0^{L'(t)} \left\{ \frac{z \sin\theta}{R^3(z')} \rho\left(z', t - \frac{R(z')}{c}\right) + \frac{z \sin\theta}{cR^2(z')} \frac{\partial \rho(z', t - R(z')/c)}{\partial t} \right\} dz'$$

$$- \hat{r} \frac{\partial}{\partial r} \left\{ \frac{Q(t - r/c)}{r} \right\} - \hat{r} \frac{\rho(L')}{R(L')} \frac{\partial L'}{\partial r} - \frac{\theta \rho(L')}{rR(L')} \frac{\partial L'}{\partial \theta} - \left(\hat{r} \cos\theta - \hat{\theta} \sin\theta\right)$$

$$\times \left[\int\limits_0^{L'(t)} \frac{1}{c^2 R(z')} \frac{\partial i\left(z', t - R(z')/c\right)}{\partial t} dz' - \left(\hat{r} \cos\theta - \hat{\theta} \sin\theta\right) \frac{i\left(L', t - \frac{R(L')}{c}\right)}{c^2 R(L')} \frac{\partial L'}{\partial t} \right]$$

Hence, we get the total electric field in spherical polar coordinate is

$$
E(r, \theta, t) = \frac{1}{4\pi \, \epsilon_0} \hat{r} \int_0^{L'(t)} \left\{ \begin{array}{c} \frac{r - z' \cos \theta}{R^3(z')} \rho\left(z', t - \frac{R(z')}{c}\right) \\ + \frac{r - z' \cos \theta}{cR^2(z')} \frac{\partial \rho(z', t - R(z')/c)}{\partial t} \end{array} \right\} dz'
$$

$$
+ \hat{\theta} \int_0^{L'(t)} \left\{ \frac{z' \sin \theta}{R^3(z')} \rho\left(z', t - \frac{R(z')}{c}\right) + \frac{z' \sin \theta}{cR^2(z')} \frac{\partial \rho(z', t - R(z')/c)}{\partial t} \right\} dz'
$$

$$
- \hat{r} \int_0^{L'(t)} \frac{\cos \theta}{c^2 R(z')} \frac{\partial i\left(z', t - \frac{R(z')}{c}\right)}{\partial t} dz' + \hat{\theta} \int_0^{L'(t)} \frac{\sin \theta}{c^2 R(z')} \frac{\partial i\left(z', t - \frac{R(z')}{c}\right)}{\partial t} dz'
$$

$$
- \hat{r} \frac{\cos \theta i\left(L't, -\frac{R(L')}{c}\right)}{c^2 R(L')} \frac{dL'(t)}{dt} + \frac{\hat{\theta} \sin \theta i\left(L', t - \frac{R(L')}{c}\right)}{c^2 R(L')} \frac{dL'(t)}{dt}
$$

$$
+ \hat{r} \left[\frac{1}{r^2} Q\left(t - \frac{r}{c}\right) + \frac{1}{rc} \frac{\partial Q(t - \frac{r}{c})}{\partial t} \right] + \hat{r} \frac{1}{cR(L')} r \left(L', t - \frac{R(L')}{c}\right) \frac{dL}{dt}
$$

$$
+ \hat{\theta} \frac{\sin \theta}{c^2 R(L')} \rho\left(L't - \frac{R(L')}{c}\right) \frac{dL'(t)}{dt}
$$

where

$$
\frac{\partial}{\partial r}\left(\frac{Q}{r}\right) = -\frac{1}{r^2} Q + \frac{1}{r} \cdot \frac{\partial Q}{\partial r} = -\frac{1}{r^2} Q - \frac{1}{cr} \frac{\partial Q}{\partial t}.
$$

Similarly,

$$
\frac{\partial L'}{\partial r} = -\frac{1}{c} \frac{\partial L}{\partial t}.
$$

and

$$
\frac{\partial L'}{\partial \theta} = \frac{\partial}{\partial \theta} \left[\frac{v}{1 - (v/c)\cos \theta} \left(t - \frac{r}{c} \right) \right]
$$

$$
= V.\left(t - \frac{r}{c}\right).(-1)\left[1 - \left(\frac{v}{c}\right)\cos \theta\right]^{-2}.\left(\frac{v}{c}\sin \theta\right)
$$

$$
= -\frac{v(t - r/c)}{\left[1 - \left(\frac{v}{c}\right)\cos \theta\right]^2}\left(\frac{v}{c}\sin \theta\right)
$$

For the special case, $\theta = 90^0$ keeping in this equation for the return stroke field at the ground level. So, $\sin \theta = 1$ and $\cos \theta = 0$. Similarly, $\hat{\theta} = -z$ since, $\hat{z} = \hat{r} \cos \theta - \hat{\theta} \sin \theta$ the unit vector $\hat{r}$ becomes only horizontal, pointing away from the channel. At $z' = 0$, a perfectly conducting plane, i.e., the earth is introduced to simulate the effect.

4.8   Determination of Electric and Magnetic Fields in Terms of Current by Using the Lorentz Condition Approach

The lightning return strokes channel as shown in Fig. 4.3 can be taken as the model in which the lightning channel travels in a straight line in a vertical direction, extended with the velocity v. Let z' be the position along z-axis in a certain time t and $i(z', t)$ be the current in the lightning channel. When the time, $t = 0$, the return stroke channel starts to propagate from the point A. Let P be the point of observer, which observes the return stroke at a distance r as shown in the Fig. 4.3. Let C be the velocity of light then the time taken by this light to travel this distance is r/c. So, the retarded current for the distance dz' is $i\left(z', t - \frac{R(z')}{c}\right)$. As mentioned earlier in the previous section $L'(t)$ be the total length of the return stroke channel at the time t, observed by the observer at the point P. Adhikari (2019c) explained that the large number of dipoles in the lightning channel length dz', and hence, the electric and magnetic fields should be produced. The vector potential at the point P due to the return stroke channel starting from the point A where $z' = 0$ to $z = L'(t)$ is given by

$A(r, \theta, \tau) = \frac{\mu_0}{4\pi} \int\limits_0^{L'(t)} \frac{i\left(z', \tau - \frac{R(z')}{c}\right)}{R(z')} \hat{z} dz'$ where τ is the time variable less than the total time t.

From the Lorentz condition,

$$\nabla . A + \frac{1}{c^2} \cdot \frac{\partial \phi}{\partial t} = 0.$$

or,

$$\frac{\partial \phi}{\partial t} = -c^2 \nabla \cdot A.$$

$$\therefore \phi = -c^2 \int\limits_{r/c}^{t} \nabla \cdot A d\tau.$$

Now, the divergence of the scalar potential,

$$\nabla \cdot A = \frac{\mu_0}{4\pi} \int\limits_0^{L'(\tau)} \nabla \cdot \frac{i\left(z', \tau - \frac{R(z')}{c}\right)}{R(z')} \hat{z} dz'$$

$$= \frac{\mu_0}{4\pi} \int\limits_0^{L'(\tau)} \left[\nabla\left(\frac{1}{R(z')}\right) i\left(z', \tau - \frac{R(z')}{c}\right) + \frac{1}{R(z')} \nabla i\left(z', \tau - \frac{R(z')}{c}\right) \right] \hat{z} dz'$$

We know, from the Fig. 4.3, $R(z') = \sqrt{r^2 + z'^2 - 2rz' \cos\theta}$.

$$\frac{dR(z')}{dr} = \frac{r - z' \cos \theta}{R(z')}$$

hence

$$\frac{dR(z')}{d\theta} = \frac{rz' \sin \theta}{R(z')}.$$

We know, $\frac{\partial i}{\partial R} = -\frac{1}{c} \frac{\partial i}{\partial t}$, where $i \equiv i\left(z', t - \frac{R(z')}{c}\right)$

$$\frac{dR(z')}{dz'} = \frac{2z' - 2r \cos \theta}{2R(z)} = \frac{z' - r \cos \theta}{R(z')}$$

$$\nabla\left(\frac{1}{R}\right) = -\frac{1}{R^2} \cdot \frac{\partial R}{\partial L} = -\frac{1}{R^2} \cdot \frac{z' - r \cos \theta}{R(z')}$$

$$= -\frac{z' - r \cos \theta}{R^3(z')}$$

$$\nabla i \equiv \frac{\partial i}{\partial z} = \frac{\partial i}{\partial R} \cdot \frac{\partial R}{\partial z} = -\frac{1}{c} \cdot \frac{\partial i}{\partial t} \times \frac{z' - r \cos \theta}{R(z')}$$

On substituting in the above equation

$$\nabla \cdot A = \frac{\mu_0}{4\pi} \int_0^{L'(\tau)} \left\{ -\left[\frac{z' - r \cos \theta}{R^3(z')}\right] i\left(z', \tau - \frac{R(z')}{c}\right) + \frac{1}{R(z')} \frac{z' - r \cos \theta}{R(z')} \right.$$

$$\times \left(-\frac{1}{c} \frac{\partial i(z', \tau - R(z')/c)}{\partial \tau} \right) \right\} dz' + \frac{\mu_0}{4\pi} \frac{L'(\tau) - r \cos \theta}{cR^2(L')}$$

$$\times i\left(L', \tau - \frac{R(L')}{c} \frac{dL'(\tau)}{d\tau} \right)$$

$$= \frac{1}{4\pi \epsilon_0 c^2} \int_0^{L'(\tau)} \left[\frac{z' - r \cos \theta}{R^3(z')} i\left(z', \tau - \frac{R(z')}{c}\right) + \frac{z' - r \cos \theta}{cR^2(z')} \frac{\partial i(z', \tau - R(z')/c)}{\partial \tau} \right] dz'$$

$$+ \frac{1}{4\pi \epsilon_0 c^2} \frac{L'(\tau) - r \cos \theta}{cR^2(L')} i$$

So, on substituting,

$$\phi = -c^2 \int_{r/c}^{t} (\nabla \cdot A) \delta \tau = \int_{r/c}^{t} \left[\frac{1}{4\pi \epsilon_0} \int_0^{L'(\tau)} \left\{ \begin{array}{l} \dfrac{z' - r \cos \theta}{R^3(z')} i\left(z', \tau - \dfrac{R(z')}{c}\right) \\[2mm] + \dfrac{z' - r \cos \theta}{cR^2(z')} \dfrac{\partial i}{\partial t}\left(z', \tau - \dfrac{R(z')}{c}\right) \end{array} \right\} dz' \right]$$

$$+ \frac{1}{4\pi \, \epsilon_0} \frac{L'(\tau) - r\cos\theta}{cR^2(L')} i\left(L', \tau - \frac{R(L')}{c}\right) \frac{dL'(\tau)}{d\tau}\Bigg] d\tau$$

On interchanging the order of integration, and as the time changing from r/c to t, the channel length $L'(\tau)$ also changing from, 0 to $L'(t)$.

We know, the time $t = \frac{L'(\tau)}{v} + \frac{R[L'(\tau)]}{c} = \frac{z'}{v} + \frac{R(z')}{c}$.

Substituting in the above equation and keeping the limit of integration then we get an expression for scalar potential as,

$$\phi = -\frac{1}{4\pi \, \epsilon_0} \int_0^{L'(t)} \left[\frac{z' - r\cos\theta}{R^3(z')} \int_{\frac{z'}{v} + \frac{R(z')}{c}}^t i\left(z', t - \frac{R(z')}{c}\right) d\tau \right.$$

$$\left. + \frac{z' - r\cos\theta}{cR^2(z')} i\left(z', t - \frac{R(z')}{c}\right) \right] dz'$$

Now, the gradient of scalar potential, $\nabla\phi$, and the time derivative of the vector potential, $\frac{\partial A}{\partial t}$.

Substituting on the equation,

$$E = -\nabla\phi - \frac{\partial A}{\partial t}$$

then we get the equation as follows.

$$E(r, \theta, t) = \frac{\hat{r}}{4\pi \, \epsilon_0} \int_0^{L'(t)} \frac{\cos\theta - 3\left\{-\frac{(z' - r\cos\theta)}{R(z')} \cdot \frac{r - z'\cos\theta}{R(z')}\right\}}{R^3(z')} \int_{t_b}^t i\left(z', \tau - \frac{R(z')}{c}\right) d\tau\, dz'$$

$$- \frac{1}{4\pi \, \epsilon_0} \hat{r} \int_0^{L'(t)} \frac{\cos\theta - 3\left\{-\frac{z' - r\cos\theta}{R(z')} \cdot \frac{r - z'\cos\theta}{R(z')}\right\}}{cR^2(z')} i\left(z', t - \frac{R(z')}{c}\right) dz'$$

$$- \frac{1}{4\pi \, \epsilon_0} \hat{r} \int_0^{L'(t)} \frac{\cos\theta + \frac{(z' - r\cos\theta)(r - z'\cos\theta)}{R^2(z')}}{cR^2(z')} \frac{\partial i\left(z', t - \frac{R(z')}{c}\right)}{\partial t} dz'$$

$$+ \frac{1}{4\pi \, \epsilon_0} \hat{\theta} \int_0^{L'(t)} \frac{\sin\theta + 3\frac{(r\cos\theta - z')(z'\sin\theta)}{R^2(z')}}{R^3(z')} \int_{t_b}^t i\left(z', \tau - \frac{R(z')}{c}\right) d\tau\, dz'$$

$$+ \frac{1}{4\pi \, \epsilon_0} \hat{\theta} \int_0^{L'(t)} \frac{\sin\theta + 3\frac{(r\cos\theta - z')(z'\sin\theta)}{R^2(z')}}{cR^2(z')} i\left(z', t - \frac{R(z')}{c}\right) dz'$$

$$+ \frac{1}{4\pi \in_0} \hat{\theta} \int_0^{L'(t)} \frac{\sin\theta + \frac{(r\cos\theta - z')(z'\sin\theta)}{R^2(z')}}{cR^2(z')} \frac{\partial i\left(z', t - \frac{R(z')}{c}\right)}{\partial t} dz'$$

$$- \frac{1}{4\pi \in_0} \hat{r} \left\{ \frac{\cos\theta - \frac{(r\cos\theta - L')}{R(L')} \cdot \frac{(r - L'\cos\theta)}{R(L')}}{c^2 R(L')} \right\} i\left(L', t - \frac{R(L')}{c}\right) \frac{dL'}{dt}$$

$$+ \frac{1}{4\pi \in_0} \hat{\theta} \left\{ \frac{\sin\theta + \frac{(r\cos\theta - L')}{R(L')} \cdot \frac{L'\sin\theta)}{R(L')}}{c^2 R(L')} \right\} i\left(L', t - \frac{R(L')}{c}\right) \frac{dL'}{dt}$$

Here, $\frac{dL'}{dt}$ is the speed of the current wave-front observed from the point P, v be the velocity of the return stroke wave-front. In the above equation, the term containing the factor $\frac{1}{R^3(z')}$ is related to static component, $\frac{1}{cR^2}$ is related to the induction component and $\frac{1}{c^2 R}$ is related to the radiation component.

References

Adhikari, P. B. (2018). Time varying electric and magnetic fields from lightning discharge. *International Journal of Electrical and Electronic Science, 5*(2), 50–55.

Adhikari, P. B. (2019a). Relation between electric and magnetic fields produced due to lightning discharge. *International Journal of Emerging Technologies and Innovative Research, 6*(5), 72–77. ISSN:2349–5162. www.jetir.org |UGC and ISSN Approved.

Adhikari, P. B. (2019b). Determination of electric field in terms of current and charge by the continuity equation approach. *International Journal of Research Granthaalayah, 7*(4), 162–170. ISSN: 2350-0530. https://doi.org/10.5281/zenodo.2653813

Adhikari, P. B. (2019c). Determination of electric and magnetic fields in terms of current by using the Lorentz condition approach. *International Journal of Engineering Research and Applications (IJERA), 9*(3), 7–10. ISSN: 2248-9622. https://doi.org/10.9790/9622-0903060710

Adhikari, P. B., & Bhandari, B. (2017). Computation of electric field from lightning discharges. International Journal of Scientific & Engineering Research, 8(9), 147. ISSN: 2229-5518.

Adhikari, P. B., & Bhandari, B, (2018a), Jefimenko equations in computation of electromagnetic fields for lightning discharges. *International Journal of Scientific & Engineering Research, 9*(6), 1678–1687. ISSN: 2229-5518.

Adhikari, P. B., & Bhandari, B., (2018b). Computation of magnetic field from lighting discharges. Journal of PMC–Physics Council 1(1), 153–157. ISSN: 2616-003X.

Cooray, V. (2003). *The lightning flash.* The Institution of Electrical Engineers (IEE)

Malan, D. J. (1963). *Physics of lightning.* English University Press.

Thottappillil, R., Uman, M., & Rakov, V. (1998). Treatment of retardation effects in calculating the radiated electromagnetic fields from the lightning discharge. *Journal of Geophysical Research, 103*, 9003–9013.

Thottappillil, R., Schoene, J., & Uman, M. (2001). Return stroke transmission line model for stroke speed near and equal that of light. *Geophysical Research Letters, 28*, 3593–3596.

Uman, M. A. (2001). The lightning discharge (p. 377). Dover Publications.

Williams, E. R. (1989). The tripole structure of thunderstorms. *Journal of Geophysical Research, 94*(13), 151–167.

Chapter 5
Measurement of Lightning

Although many types of research have addressed various aspects of lightning discharge phenomena, a complete theory does not exist. It has been investigated by different methods such as photography, current measurement, optical, electromagnetic, and acoustics, or a combination of two or three of these methods (Ogawa, 1995). Among them, the combination of electromagnetic fields and photography methods has achieved more success. Some methods are described here as an example of the measurement process.

5.1 Electric Field Measurement Method

In this method, there are several processes like horizontal parallel plate antenna to measure the vertical electric field, vertical rod antenna to measure the horizontal electric field, and so on.

5.1.1 Horizontal Parallel Plate Antenna

In 1916, Wilson in England, the third branch of research on the nature of lightning discharge phenomena, was concerned with measuring electric fields. He studied the invention of the cloud chamber to track high-energy particles. Wilson, Nobel Prize Winner, was the first member to use the electric field measurements to estimate the charge structure in the cloud. Berger (1977), Rakov and Uman (2003), Nanevicz et al. (1987) described that various motivations for studying the electrical properties of lightning and the complexity of the flash have continued to challenge the creativity of lightning experimenters and instrument designers. A downward-directed electric field is considered positive in the atmospheric sign convention, as shown in Fig. 4.1,

© The Author(s), under exclusive license to Springer Nature Singapore Pte Ltd. 2022   89
P. B. Adhikari and A. Adhikari, *Lightning Discharges*,
SpringerBriefs in Applied Sciences and Technology,
https://doi.org/10.1007/978-981-19-1926-8_5

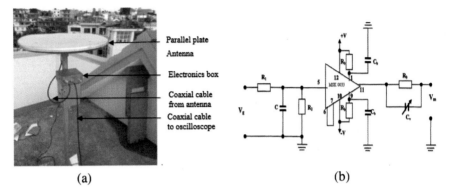

(a)   (b)

Fig. 5.1 Measurement of horizontal magnetic field in which **a** the elevated parallel plate antenna installed in Himalayan region and **b** the electronic circuit pertinent to buffer amplifier used

and an upward-directed electric field is considered negative. According to this definition, a lightning flash that transports a negative charge to the ground gives rise to a positive electric field change. A lightning flash that transports a positive charge to the ground gives rise to a negative electric field change (Adhikari et al., 2016).

Adhikari et al., (2016, 2017) explained that the electromagnetic radiations of various wavelengths of various frequencies are radiated from lightning discharge phenomena. The electric and magnetic field signatures produced due to lightning are the basic physical parameters in understanding the mechanisms of lightning discharges. These radiations due to lightning travel from the discharge channel in all possible directions, of which the horizontal parallel plate antenna senses the vertical electric field. The different field signature waveforms captured by the parallel plate antenna are connected via an electronic circuit through the coaxial cable (Adhikari, 2019). The upper plate of the antenna is connected with a buffer circuit, and the lower plate is connected to the ground. The electronic circuit with the antenna system and the buffer amplifier is shown in Fig. 5.1.

5.1.2   Basic Theory of Antenna System

The parallel plate antenna captured the vertical electric field produced due to the lightning discharge. The static electric field theory can be used if the wavelength of the electric field is larger than the size of the circular metallic plate of the antenna, as shown in Fig. 5.2a. Similarly, (b) and (c) represent that the size of the circular metallic plate of the antenna and wavelength of the electric field are comparable, and the wavelength of the electric field is very small than the size of the circular metallic plate of the antenna, respectively.

Let us consider Q be the total charge induced due to the lightning on the horizontal parallel plate. The electric field appears in the metallic plate considering the size of

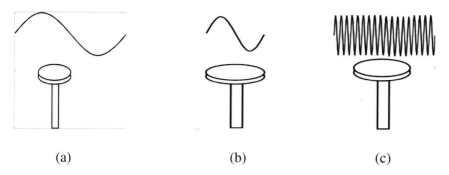

Fig. 5.2 Comparing the wavelength of the electric field with the size of antenna. In **a** wavelength is larger than the size of the antenna, in **b** equal in size and in **c** wavelength of the electric field is very smaller than the size of the antenna

the metallic plate is much smaller than the wavelength of the electric field as shown in Fig. 5.2, so, the static electric field theory can be applied. We know from the Gauss's law, the total magnitude of the charge is the integration of the total electric flux density over the surface, then

$$Q = \int_s D \cdot ds \tag{5.1}$$

where the electric flux density is given by

$$D = E \,\epsilon_0 \epsilon_r$$

where ϵ_0 and ϵ_r are the permittivity of the vacuum and medium, respectively, D is the electric flux density and E be the electric field intensity. If the electric field intensity E is uniform over the metallic plate of the parallel plate antenna, then the Eq. (5.1) becomes $Q = D \cdot S$, where S is the total surface area of the plate. Then the potential difference developed between the charged (upper) plate at a height "d" and the ground is

$$V_g = E_n \cdot d \tag{5.2}$$

If "R" be the resistance between the plate and the ground, then measured voltage is $V_g = Ri$ and the current through the resistor is $i = \frac{dQ}{dt}$. On substituting the value of Q in terms of normal electric field E_n, then we get,

$$V_g = \epsilon_0 \epsilon_r \, R \cdot S \cdot \frac{dE_n}{dt} \tag{5.3}$$

Hence, from this relation the voltage is in terms of the derivative of electric field. Again, if we consider the capacitor C then the voltage is measured across it. Hence, the voltage across the capacitor "C" is given by:

$$V_g = \frac{Q}{C} \qquad (5.4)$$

Now, on substituting the value of Q in terms of normal electric fields E_n, then the voltage across the capacitor is

$$V_g = \frac{\epsilon_0 \epsilon_r}{C} \frac{S}{}.E_n \qquad (5.5)$$

Hence, the measured voltage across the combination of capacitor and a resister depends on the normal electric field and derivative of normal electric field. In this case, the measured voltage (V_m) is less than V_g as the RC circuit as shown in Fig. 5.1. The value of R is chosen such that the value of voltage across C is so, small on compared with the voltage across R, then the effect of capacitor only implemented. The measured voltage across RC circuit is:

$$V_m = E_n.d \cdot \frac{C_g}{C_g + C} = V_g \frac{C_g}{C_g + C}$$

where C_g is the capacitance of the upper plate of the antenna with respect to ground. The values of components are in the Fig. 5.1 above are as follows: $R_1 = 50\ \Omega$; $C = 15$ pF; $R_2 = 99$ MΩ; $C_b = 0.1\ \mu$F; $C_v = 10$–60 pF; $R_0 = 43\ \Omega$, the output resistance of the buffer amplifier is 7 Ω. Hence, the total output resistance becomes 50 Ω. By substituting these values, then the decay time constant for the parallel plate antenna is 13.4 ms, which is large enough to record the radiation components of electric field produced by lightning. Then the measured voltage is

$$V_m = E_n.d_{\text{eff}} \cdot \frac{C_g}{C_g + C_c + C}.$$

where C_c is the capacitance of the coaxial cable used to transmit the field from antenna to the recorder (Adhikari, 2019; Ahmad et al., 2010; Johari et al., 2016; Baharudin et al., 2012; Galvan & Fernando, 2000; Sharma et al., Sharma et al. 2008a, b). Digital Storage Oscilloscope (DSO) or Pico-scope 6404D is a special type of oscilloscope which digitizes the analogue voltage and store the signal digitally.

5.1.3   Vertical Rod Antenna

The vertical rod antenna has been employed to sense the slow electrostatic field change due to close lightning discharge. A schematic diagram of a vertical rod antenna is given in the following Fig. 5.3 in which the height of the upper metallic

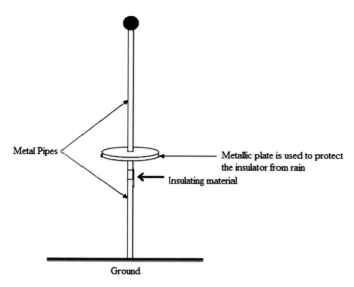

Fig. 5.3 Vertical rod antenna to measure the horizontal electric field (Sharma, 2007)

pole is 3.3 m and that of the lower one is 1.45 m. The two poles are insulated from each other using an insulator 5 cm thick. A metallic plate is used to protect the insulator from rain. Since rainwater may disturb the insulation, the metallic plate is used to protect the insulator from rainwater (Sharma, 2007). Electrically conducting ball at the top of the antenna is used to prevent possible corona emission as in the figure. Kannangara et al. (1978) have calibrated the antenna and obtained effective height of the antenna system as 1.88 m. Except for the height of the antenna; other parameters of antenna system are same as that of the parallel plate antenna explained above. The signal record the slow field variation the decay time constant had to be increased. For this purpose, the capacitance of the RC circuit has been increased to 10 nF; as a result, the value of decay time is obtained as 1 s which is large enough to faithfully measure the slow field variation. For the given values of circuit components, the electric field impinging the antenna can be given as:

$$E_n = 45.8 \times V_m$$

5.1.4   Narrow Band Electric Field Measurement System

Apart from the broadband electric field measurement system, narrowband measuring systems have also been employed in this study. The narrowband system is simply an LCR circuit tuned at a particular frequency. It consists of an inductor which, along with the antenna capacitance and cable resistance, forms a tuned LCR circuit. This

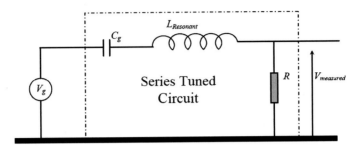

Fig. 5.4 An equivalent circuit of the HF narrowband receiver formed together combined with antenna capacitance and cable resistance (Sharma, 2007)

system has been employed to investigate the high-frequency content of the different events of the lightning flash. Two of such systems tuned at 5 and 10 MHz have been used. The bandwidth of these narrowband receivers has been obtained as 471 and 2020 kHz, respectively. A schematic diagram of the narrowband receiver has been depicted in Fig. 5.4. The value of antenna capacitance (Cg) is 59 pF, the cable resistance (R) is 50 Ω, and the inductance of the coil for 5 MHz is 17 μH, and that for 10 MHz is 4.29 μH (Edirisinghe et al., 2006).

5.2 Photography Method

In 1889, Hoffert in England and Weber and Walter in Germany used the photography method as the first method of measuring lightning, after Benjamin Franklin, in 1752, proved that the lightning phenomenon is a discharge phenomenon. For the photography method, it was difficult before the invention of the streak camera. The streak camera is an instrument in which there is a motion of the lens with the film continuously. Then Boys invented the two-lens streak camera, also called the Boys camera, and was used by McEachron. Schonland and Collens (1934), Schonland et al. (1935) also used the earlier version of this camera. However, Idone and Orville (1982), Jordan and Uman (1983) used the modern version of the Boys camera modified by Backman and Whitley. In the revised version, Backman and Whitley gave the model 318 streak camera, whose time resolution is in the order of 1 μs. Orville, in 1968, used the instrument spectrometer, which measures the spectrum recorded on a streak camera (Uman, 2001).

5.3 Thunder Measurement Method

In this method, the sound produced during lightning discharge phenomena called thunder is captured by the microphone. The microphones give the signals of the thunder (sound), which are recorded in a tape-recorder. Bohannon, Balchandran, etc. used the low-frequency microphone of 0.1–300 Hz, but Holms used a capacitor microphone of wide frequency of 0.3–20 kHz (Uman, 2001).

5.4 Current Measurement Method

In 1897, Pockels in Germany used the current measurement method as a second process in the research field to know about the phenomena of lightning discharge. He began the development of magnetic links by measuring the amplitudes of lightning current (Berger, 1977; Sharma, 2007; Uman, 2001).

5.5 Magnetic Field Measurement System

The measurement of the magnetic field produced by lightning can be dated from 1897 when Pockels observed that the residual magnetism induced in a piece of nepheline basalt by a unidirectional magnetic field depended neither on the duration nor the time variation of the field but only on its maximum value. Several researchers have made use of magnetizable materials to investigate the lightning channel current. The majority of them being employed in the vicinity of tall structures, the probability of being struck by lightning is maximum. Used a loop antenna to estimate the current in a lightning channel striking on open terrain. Krider et al. (1975) designed a broadband antenna system using a conducting loop and an active integrator. This system has been widely employed in the lightning magnetic field measurement system and lightning locating system. Although anisotropic magneto-resistive (AMR) sensors have been used in navigation, vehicular detection, etc., they have apparently not been used in the lightning measurement system. AMR sensors are small solid-state devices with fast response and reasonably wide bandwidth (Sharma, 2007) (Fig. 5.5).

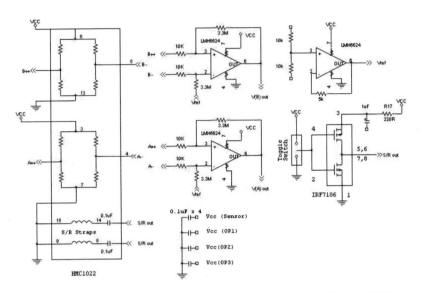

Fig. 5.5 Block diagram of the electronic circuit for HMC1022 sensor (Sharma, 2007)

References

Adhikari, P. B., Sharma, S. R., & Baral, K. N. (2016). Features of positive ground flashes observed in Kathmandu, Nepal. *Journal of Atmospheric and Solar-Terrestrial Physics, 145*, 106–113.

Adhikari, P. B., Sharma, S. R., Baral, K. N., & Rakov, V. A. (2017). Unusual lightning electric field waveforms observed in Kathmandu, Nepal, and Uppsala, Sweden. *Journal of Atmospheric and Solar-Terrestrial Physics, 164*, 172–184.

Adhikari, P. B. (2019). Measurement of electric fields due to lightning radiation (Vol. 905, pp. 38–43). AISC, Springer Nature Switzerland AG. https://doi.org/10.1007/978-3-030-14680-1_5

Ahmad, N. A., Fernando, M., Baharudin, Z. A., Cooray, V., Ahmad, H., & Malek, Z. A. (2010). The characteristics of narrow bipolar pulses in Malaysia. *Journal of Atmospheric and Solar-Terrestrial Physics, 72*, 530–540.

Baharudin, Z. A., Fernando, M., Ahmad, N. A., Makela, J. S., Rahman, M., & Cooray, V. (2012). Electric field changes generated by the preliminary breakdown for the negative cloud-to-ground lightning flashes in Malaysia and Sweden. *Journal of Atmospheric and Solar-Terrestrial Physics, 84*, 15–24.

Berger, K. (1977). *The earth flash, in lightning, physics of lightning.* Academic Press.

Edirisinghe, M., Makela, J. S., Montano, R., Fernando, M., & Cooray, V. (2006). *28th International conferences on lightning protection Kanazawa, Japan* (pp. 1–17).

Galvan, A., & Fernando, M. (2000). *Operative characteristics of a parallel-plate antenna to measure vertical electric fields from lightning flashes.* Uppsala University.

Idone, V. P., & Orville, R. E. (1982). Lightning return stroke velocities in the thunderstorm research international program (trip). *Journal of Geophysical Research, 87*, 4903–4915.

Johari, D., Cooray, V., Rahman, M., Hettiarachchi, P., & Ismail, M. (2016). Characteristics of leader pulses in positive ground flashes in Sweden. *Electric Power System Research, 153*, 3–9.

Jordan, D. M., & Uman, M. A. (1983). Variation in light intensity with height and time from subsequent lightning return stroke. *Journal of Geophysical Research, 88*, 6555–6562.

Kannangara, M. L. T., Lundquist, S., & Pisler, E. (1978). The effective height of vertical arial of a lightning flash counter (UURIE:106–178). Institute for High Voltage, Uppsala University.

Krider, E. P., Radda, G. J., & Noggle, R. C. (1975). Regular radiation field pulses produced by intra-cloud lightning discharges. *Journal of Geophysical Research, 80*, 3801–3804.

Nanevicz, J. E., Vance, E. F., & Hamm, J. M. (1987). Observation of lightning in the frequency and time domains. *Lightning Electromagnetics, 7*, 267–286.

Ogawa, T. (1995). *Lightning currents, handbook of atmospheric electrodynamics.* CRC.

Rakov, V. A., & Uman, M. A. (2003). *Lightning: Physics and effects.* Cambridge University Press.

Schonland, B. F. J., & Collens, H. (1934). Progressive lightning. *Proceedings of Royal Society, A143*, 654–674.

Schonland, B. F. J., Malan, D. J., & Collens, H. (1935). Progressive lightning ii. *Proceedings of Royal Society, A152*, 595–625.

Sharma, S. (2007). *Electromagnetic fields radiated by lightning in tropical and temperate regions.* Ph.D. Thesis. Faculty of Science, University of Colombo, Sri Lanka

Sharma, S. R., Cooray, V., & Fernando, M. (2008a). Isolated breakdown activity in swedish lightning. *Journal of Atmospheric and Solar-Terrestrial Physics, 70*, 1213–1221.

Sharma, S. R., Fernando, M., & Cooray, V. (2008b). Narrow positive bipolar radiations from lightning observed in Sri Lanka. *Journal of Atmospheric and Solar-Terrestrial Physics, 70*, 1251–1260.

Uman, M. A. (2001). *The lightning discharge* (p. 377). Dover Publications.

Printed in the United States
by Baker & Taylor Publisher Services